Disclaimer

The publisher of this book is by no way associated with the National Institute of Standards and Technology (NIST). The NIST did not publish this book. It was published by 50 page publications under the public domain license.

50 Page Publications

Book Title: An Independent Measurement System for Performance Evaluation of Road Departure Crash Warning Systems

Book Author: Sandor S. Szabo; Richard J. Norcross

Book Abstract: The Independent Measurement System (IMS) is a general-purpose measurement system designed for collecting and analyzing data needed for evaluation of intelligent vehicle systems. The IMS consists of sensors, data collection hardware and analysis software independent of the system under test. This provides objectivity, quality assurance and redundancy to data collection and testing activities. The following version of the IMS supports testing and performance evaluation of the road departure crash warning system (RDCWS) developed for a Department of Transportation Field Operation Test under the Intelligent Vehicle Initiative.

Citation: NIST Interagency/Internal Report (NISTIR) - 7287

Keyword: Intelligent Vehicle Initiative;performance measurement;road departure;safety system;test procedures;warning system

NISTIR 7287

An Independent Measurement System for Performance Evaluation of Road Departure Crash Warning Systems

Sandor Szabo
Richard Norcross
U.S DEPARTMENT OF COMMERCE
Technology Administration
National Institute of Standards and Technology
Intelligent Systems Division
Gaithersburg, MD 20899-8230

National Institute of Standards and Technology
Technology Administration, U.S. Department of Commerce

NISTIR 7287

An Independent Measurement System for Performance Evaluation of Road Departure Crash Warning Systems

Sandor Szabo
Richard Norcross
U.S DEPARTMENT OF COMMERCE
Technology Administration
National Institute of Standards and Technology
Intelligent Systems Division
Gaithersburg, MD 20899-8230

January 2006

U.S. DEPARTMENT OF COMMERCE
Carlos M. Gutierrez, Secretary
NATIONAL INSTITUTE OF STANDARDS AND TECHNOLOGY
William Jeffrey, Director

An Independent Measurement System for Performance Evaluation of Road Departure Crash Warning Systems

January 3, 2006

Prepared By:
Sandor Szabo and Richard Norcross
National Institute of Standards and Technology
100 Bureau Drive, Mail Stop 8230
Gaithersburg, MD 20899-8230

This document was prepared for the U.S. Department of Transportation under interagency agreement DTFH61-00-Y-30132. Send comments and/or questions to sandor.szabo@nist.gov.

Certain commercial entities, equipment, or materials may be identified in this document in order to describe an experimental procedure or concept adequately. Such identification is not intended to imply recommendation or endorsement by the National Institute of Standards and Technology, nor is it intended to imply that the entities, materials, or equipment are necessarily the best available for the purpose.

This page blank for two-side printings.

Table of Contents

1 Introduction..1
2 Data Collection System..3
3 Camera Calibration Setup..5
   3.1 Side View Calibration Sticks..6
   3.2 Front View Calibration Sticks...6
4 Camera Calibration..8
   4.1 Creating the Calibration Video...8
   4.2 Using the Calibration Program...8
      4.2.1 Load Calibration Image..9
      4.2.2 Crop Calibration Image...10
      4.2.3 User define cal points...12
      4.2.4 Learn Calibration..15
      4.2.5 Check Calibration...15
      4.2.6 Save Calibration..15
      4.2.7 Return..15
   4.3 Considerations for Calibrating the Front View...15
   4.4 Creating an INI File for the Video Calibration...16
   4.5 Calibration Verification...17
5 Data Processing...19
   5.1 Time codes..19
   5.2 Video and Data Timing...21
   5.3 Creating the avi file using Adobe Premier..22
   5.4 Creating the INI file..26
   5.5 Creating the VMS (GPS) file..27
6 AnalyzeData Main Control Panel..28
   6.1 Starting and Stopping..28
   6.2 Video Frame Controls...29
   6.3 Lateral Measurements...29
   6.4 Lane Width and Vehicle Offset...29
   6.5 Longitudinal Measurements...30
   6.6 GPS Data..31
   6.7 Control Panels..31
7 Locating Video Based on GPS Time...32
8 Lateral Drift Analysis..34
   8.1 Overview..34
      8.1.1 Warning Time...34
      8.1.2 Weather Environment...34
      8.1.3 Warning Type..35
      8.1.4 Road Type...35
         8.1.4.1 Marker Condition..35
         8.1.4.2 Marker Color..35
         8.1.4.3 Marker Type...35
         8.1.4.4 Curve Direction...35
         8.1.4.5 Curve Entry/Exit...36

        8.1.4.6 Freeway Indication 36
        8.1.4.7 AMR 36
        8.1.4.8 AMR Type 36
    8.1.5 Vehicle's Lane Information 36
        8.1.5.1 Lane Position 37
        8.1.5.2 Lateral Velocity 37
        8.1.5.3 Discrete Obstacles 37
    8.1.6 Warning Classifications 38
        8.1.6.1 Scenario Identification 38
        8.1.6.2 Challenging Environment 38
        8.1.6.3 Alert Ratings 39
  8.2 Positive Alert Measurement Procedure 39
  8.3 Negative Alert Measurement Procedure 42
  8.4 Atmospheric/Cloud Indication 42
9 Lane Marker Contrast 45
10 MapPoint Map Display 48
References 50
Appendix A Adobe Premier Notes 51
Appendix B Creating a DVD using DVDit! 52
Appendix C Extracting GPS data from video using VMS and MediaMapper 56
Appendix D Nonlinear Camera Calibration 62
Appendix E UTM Conversions 64

1 Introduction

The Independent Measurement System (IMS) is a general-purpose measurement system developed by the National Institute of Standards and Technology (NIST) for collecting and analyzing data needed for evaluation of intelligent vehicle systems. The IMS consists of sensors, data collection hardware and analysis software independent of the system under test. This provides objectivity, quality assurance and redundancy to data collection and testing activities. The following version of the IMS supports testing and performance evaluation of the road departure crash warning system (RDCWS) developed for a Department of Transportation Field Operation Test (FOT) under the Intelligent Vehicle Initiative [1].

In the RDCWS program Phase 1 verification test, NIST used the IMS to evaluate the RDCWS on a series of track-based tests. The program proceeded with Phase 2 only after successful verification tests. The IMS also supported a Volpe National Transportation Systems Center task to characterize warning system performance in on-road real-world conditions. NIST collected and processed data and performed analysis of lateral drift warning scenarios.

This document describes the current IMS version and contains the following information:

Section 2 describes the IMS hardware (a four-camera system with video multiplexor with digitally inserted DGPS and audio recorded on a digital video recorder) and installation.

Section 3 describes the visual targets for camera calibration and their proper use.

Section 4 describes the camera calibration process.

Section 5 describes how evaluators process the data to produce the final video, GPS and support files.

Section 6 describes the data analysis tools including controls and displays for scrolling through the video, plotting vehicle trajectory, viewing time and making a variety of road measurements.

Section 7 describes the procedure to locate video frames taken at a specific GPS time. This procedure supports analysis of warning system data collected by the University of Michigan Transportation Research Institute data collection system (DAS).

Section 0 describes the lateral drift warning analysis procedure used to support Volpe's on-road characterization of the RDCWS.

Section 9 describes a tool for measuring lane marker contrast.

Section 10 describes how to display IMS data on a MapPoint map.

The analysis software is written in LabView. An executable-only version of the software is available at ftp://ftp.nist.gov/pub/mel/szabo/AnalyzeDataDistribution/. The software requires a run-time license from LabView. Contact Sandor Szabo (sandor.szabo@nist.gov) to obtain the license.

2 Data Collection System

This section describes the data collection system used during the RDCWS FOT verification tests from May 2003 until March 2005.

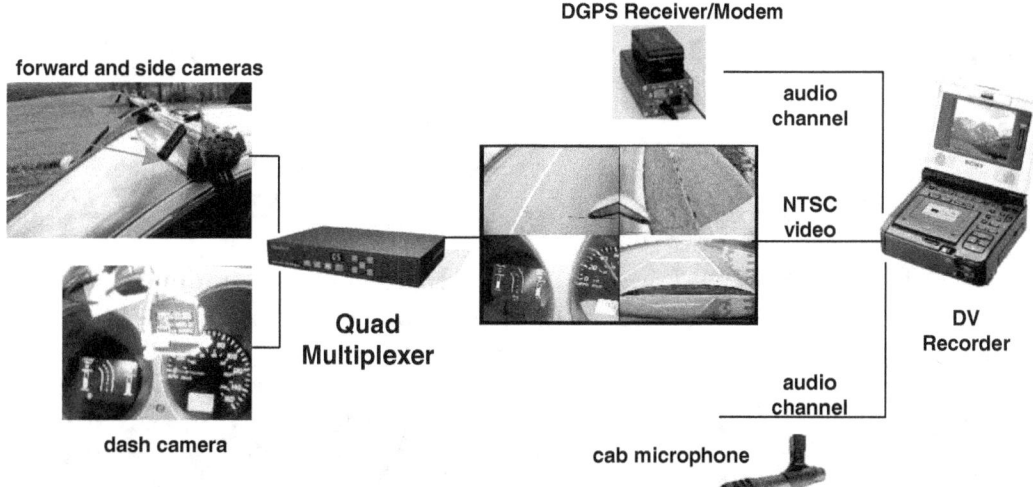

Figure 1. System diagram of data collection system.

The on-board data collection system (see Figure 1) consists of four cameras, a quad multiplexer, a microphone, a differential GPS (DGPS) receiver/modem and a digital video tape deck. Three cameras mount to a removable roof rack. Two roof cameras view the lanes to the right and left of the front wheels. These cameras have infrared LEDs for nighttime illumination. The center camera records a view similar to the warning system view. The fourth camera mounts to the dash and views the warning system display and the speedometer (useful if GPS data is not available). A microphone in the cab captures the audible warnings and voice feedback from the driver.

The dash camera and audible alerts pinpoint, from the perspective of the driver, an alert event. Figure 2 through Figure 5 show cautionary and imminent lateral drift and curve speed warnings. In these figures, the upper left and right quads show the left and right side of the vehicle (centered on front wheels) respectively. The lower right quad shows the view in front of the vehicle. The lower left quad shows the warning system display and the dash speedometer. The IMS multiplexes the quad images with the GPS data and audio in real-time. Thus, the data suffers minimal errors due to misalignment of data streams. The recorded data correlates the vehicle position and motion with the warning system events.

The box shown in Figure 6 contains the data collection system's electronic components. A VMS-200 DGPS receiver/modem interfaces with a roof-mounted GPS antenna and the digital video (DV) recorder. The VMS outputs satellite data as an audio signal to one of the DV recorder's audio channels. This function is similar to overlaying GPS data graphically on the video signal, except the "overlay" is in a computer readable format. Later, a suite of software extracts the GPS data from the tape using the VMS, performs differential correction to the GPS data, and stores the GPS locations into a file. When viewing a video file, the analysis software uses the tape time to search for the DGPS location that corresponds to a particular video frame.

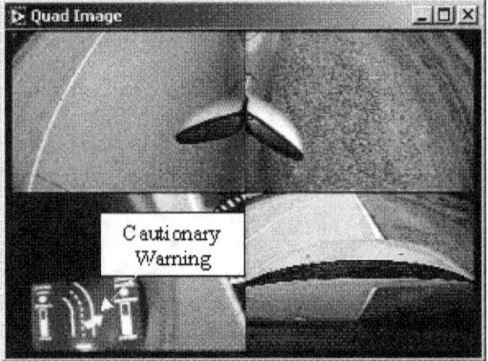

Figure 2. Lateral drift cautionary warning indicated by arrow at 2 o'clock.

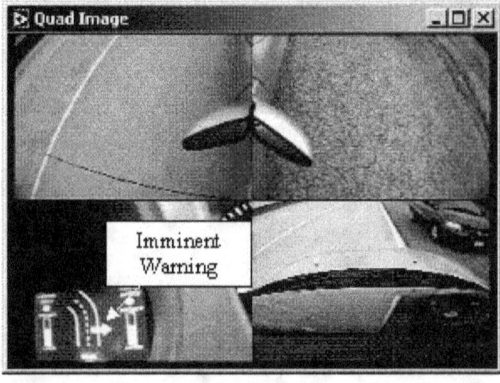

Figure 3. Lateral drift imminent warning indicated by arrow at 3 o'clock.

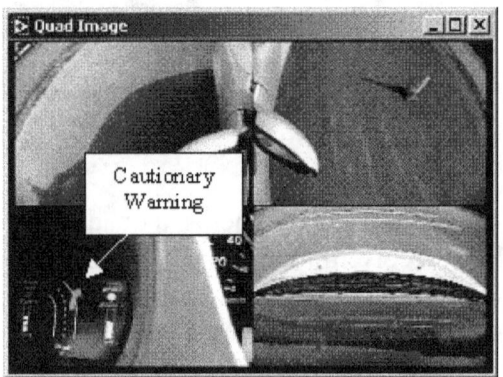

Figure 4. Curve speed cautionary warning indicated by short arrow.

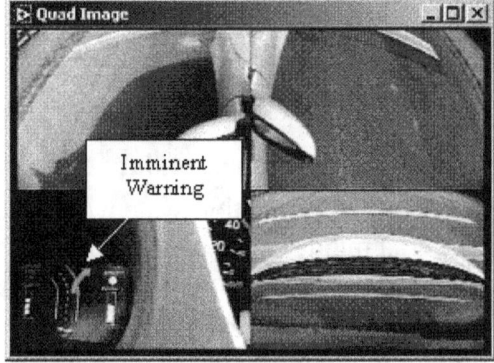

Figure 5. Curve speed imminent warning indicated by long arrow.

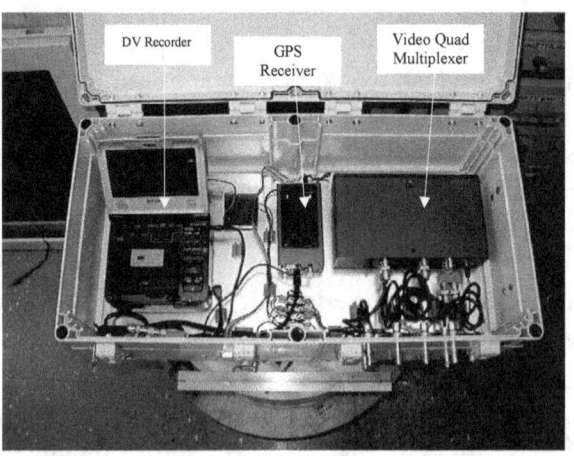

Figure 6. The electronics box installed in the vehicle for data collection.

3 Camera Calibration Setup

Camera spatial calibration defines the transformation between image points (i.e., pixel coordinates) and world points on the road surface surrounding the vehicle. This section describes the data collection procedures for the calibration process. This section primarily concerns the data collectors. Data users may skim this section. However, users should be aware of how to check the calibration as described in Section 4.5 Calibration Verification.

NIST implemented two forms of calibration techniques for the RDCWS FOT. A non-linear calibration technique took measurements over a large area (see Appendix D). However, the large grid system required careful assembly and handling. Although measurements on the grid points were accurate, the measurements *between* grid points suffered distortion.

A simpler calibration scheme supports measurements along a single line on the ground. The line is a series of meter sticks (called cal sticks). The sticks have alternating black and white segments. The transition between black and white is a *control point* for calibration purposes. The widths of the segments vary to compensate for distortions in the image. Narrow segment sticks lie closer to the camera and wide segment sticks lie farther from the camera. Figure 7 shows the calibration sticks for the right side camera. The calibration uses four 1-meter sticks. The first two (starting from closest to the vehicle on the right side of the image) have 10 cm segments. The third and fourth sticks have 25 cm and 50 cm segments, respectively. Note in Figure 7 how the 50 cm segments appear the same width as the closest 10 cm segments. The forward calibration (Figure 8) uses six alternating black and white 1-meter sticks.

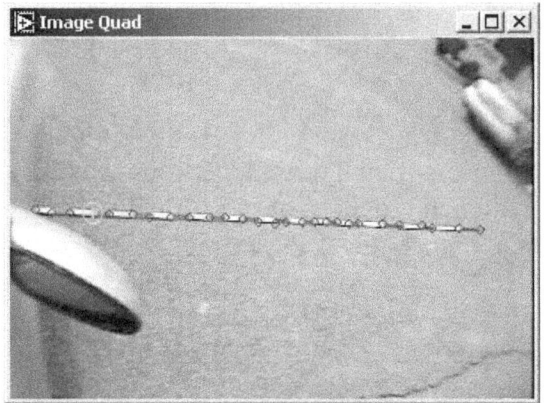

Figure 7. Calibration sticks for side view camera.

Figure 8. Forward camera calibration using one-meter sticks.

The calibration procedure identifies the pixels corresponding to the control points (i.e., segments' endpoints). The computer software displays the control point pixels with red circles. The first circle (furthest left) correlates to 0 m from the right wheel. The last red circle (furthest right) correlates to the combined lengths of the cal sticks or 4 m. Measurements in the image are valid only along the red circles. A linear interpolation generates the measurement value between the circles.

3.1 Side View Calibration Sticks

The following steps describe the camera and calibration stick set-up for the side view cameras on the Nissan Altima used in the FOT.

1. Place 4 1-meter sticks end-to-end in line parallel to the front axle and adjacent to the front wheel (see Figure 7 for right side of vehicle). Use the sticks with alternating 10 cm black and white segments for the two closest to the wheel. For the third and fourth sticks use the stick marked with 25 cm segments and 50 cm segments, respectively.
2. Center the sticks vertically in the camera image. When centered, the sticks appear straight as opposed to parabolic (which is due to the radial distortion of the lens).
3. Align the inside stick with camera image edge. Ensure the image includes the vehicle door and the mirror. The stick alignment produces ideal measurements close to the vehicle and the mirror provides a landmark to detect if the camera moves. (Note: for other vehicles, one may need to add a visual landmark.)
4. Using a video display to view the camera output, mark the location of the mirror (or other visual landmark) using a fine pen to trace the mirror's contour. If the camera moves later, the location of the mirror in the display will not align with the original tracing. Before collecting video data, verify the alignment of the mirror with the tracing. In most cases, if the camera moves the user can realign the camera to match the original tracing. If not, the user must lay down meter sticks and perform another calibration.
5. Record several seconds of the calibration sticks.
6. Rewind the video and verify recording of calibration sticks.
7. Repeat Steps 1 through 6 for the opposite (in this case right) side view.

3.2 Front View Calibration Sticks

The following steps describe the camera and calibration stick set-up for the front view camera on the Nissan Altima used in the FOT. One significant difference from the side view cameras is the need to determine the translation and rotation of the sticks with respect to the vehicle. Another need is to have someone point to the control points since they may be difficult to identify in the image.

1. Place six sticks in front of the vehicle in a straight line. Alternate a black meter and a white meter stick (the entire stick is the same color, not segmented). The sticks should lie just above the hood in the image (roughly 2 m in front). Place the sticks perpendicular to the vehicle heading and align the center of the six sticks with the center of the vehicle.
2. Measure the distance from the rear wheel to the edge of the sticks (line C in Figure 9), the distance from the rear wheel to the stick closest to the front wheel (line B in Figure 9), and the distance from the intersection of line B with the sticks to the edge of the sticks (line A in Figure 9. The calibration process uses the three values to

determine the position and orientation of the cal sticks with respect to the vehicle's front wheels. (Note that if not known already, the calibration requires the distance between the rear-wheel and the front wheel. Update the value if using a car other than the Nissan Altima.)

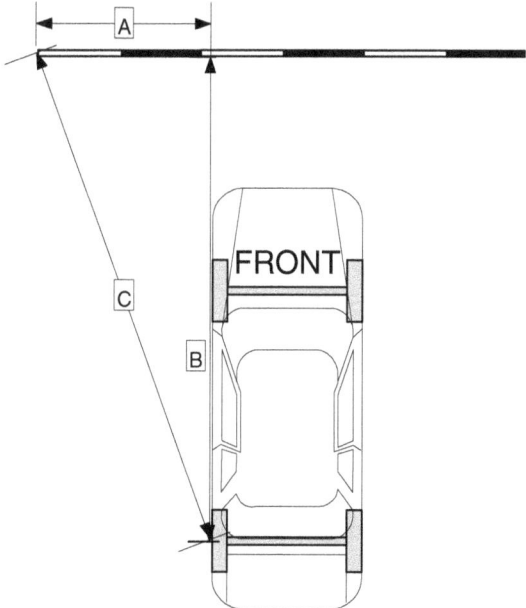

Figure 9. Three measurements to determine offset and rotation of front calibration stick with respect to the vehicle.

3. Manually adjust the camera to vertically center the sticks in the video image. When centered the sticks appear straight as opposed to parabolic (which is due to the radial distortion of the lens). The correct position includes the vehicle's hood and minimizes the sky (reduced sky limits image saturation from the sun).
4. Horizontally center the sticks in the camera image.
5. Using an external video display showing the camera's output (the lower right quad), cover the forward image of the car's hood with clear tape. Using a fine pen, outline (trace) the perimeter of the hood. Before collecting video data, verify alignment by comparing the hood with the traced perimeter. Realign the camera if the image does not align with the contour.
6. Turn on video.
7. Use a pointer (e.g., the white or black meter stick) and point out each control point (walk to each point, point at the control point and maintain for 2 seconds).
8. Rewind the video and verify that the tape contains the calibration sticks and that the pointer stick is visible. It may be hard to see the bottom of the pointer stick depending on the background and lighting. You may have to experiment a few times or wait for better lighting conditions.

4 Camera Calibration

The procedures in Section 3 generate a videotape of the calibration sticks at specific places around the vehicle. The procedures in this section extract data from the videotape and create a calibration file. The calibration file maps the quad image pixels to the GPS coordinates and the distance offset of the pixel from either front wheel.

4.1 Creating the Calibration Video

The calibration procedures use video files in the avi format. Several commercial software packages can translate a video image into an avi file. Section 5.3 Creating the avi file using Adobe Premier explains one such package. The avi file may be from a single video file or may be spliced from several video files but must contain video of all three sets of calibration sticks (left, right, and forward).

4.2 Using the Calibration Program

A program, written by NIST researchers, assists the extraction of calibration data from the avi file of the calibration sticks. The program is a LabView virtual instrument (vi) called "Nonlinear Calibration.vi". The program produces the video calibration file for the cameras in the current position and orientation. If the camera view changes, i.e., the pixels do not align with the original pixels covering the calibration stick, then a new calibration file must be produced.

Figure 10 shows the vi's control panel. The buttons on the left, labeled "Calibration Steps" step through the calibration process. The user must execute the steps in order from top to bottom. The user cannot skip steps nor return to a previous step. If an error occurs, the user starts over by using the "Return" button. The following sections describe each step in the calibration process.

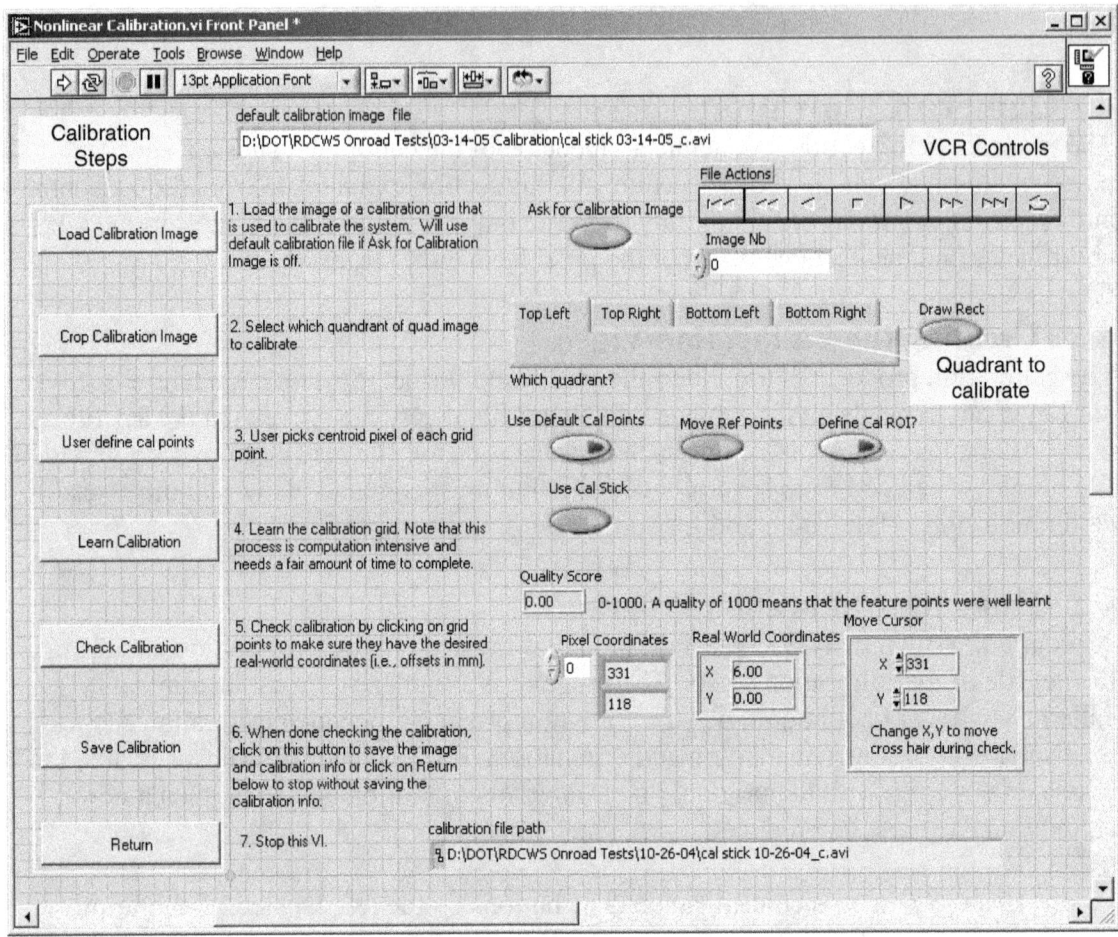

Figure 10. Control Panel for Calibrating Video.

4.2.1 Load Calibration Image

Click on the "Load Calibration Image" button to load the avi file containing the calibration stick images. When "Ask for Calibration Image" is off (green area dark) the program uses the file listed in "default calibration image file". Otherwise, the program displays a file-load dialog box and loads the new file into "default calibration image file". To skip going through the file-load dialog box after calibrating the first quad (the process requires reloading the image file to calibrate each quad), turn "Ask Calibration Image" on for the first quad, and off for the remaining quads. The program uses the same avi file in the subsequent quads. When the image loads, a separate window appears with the quad image avi (see Figure 11).

Figure 11. The first frame of the calibration avi file.

4.2.2 Crop Calibration Image

The "Crop Calibration Image" button initiates a separate vi for interactively specifying the coordinates of one of the four quadrants. Before using the "Crop Calibration Image" button, the user selects the quadrant to calibrate. With the VCR controls ("File Action"), the user advances the avi file to a frame that contains the calibration sticks. The frame in Figure 11, shows the calibration sticks in the right camera view. The remainder of this section describes the calibration of right side view in the top right quadrant.

"Which quadrant?" is a radio button switch. To calibrate the top right quadrant, the user selects "Top Right". (While the top left is similar, the bottom right quadrant has different steps described later.)

The "Draw Rect" button initiates an interactive box to define the region of the image that corresponds to the top right. When this button is off, the routine uses the "Top Right Default" coordinates that appear off the screen to the right. The routine saves this value when the user turns off "Draw Rect". If an error requires a re-run, the user can re-use the value without redrawing the box.

Click on "Crop Calibration Image" to define the quadrant. The image appears in a window as shown in Figure 12. Use the mouse to draw a rectangle around the quadrant. (Note: Figure 12 shows a box around the top right quadrant.) Adjust the box with the handles at the box's corners. Click on OK when finished. A dialog box asks if you would like to reset the coordinates for this quad. Click yes and the routine changes the default values to the current values. Click no if there was some type of error and you do not wish to save these values. Afterwards, the top right quadrant appears in a separate window (Figure 13).

Figure 12. The crop image control panel.

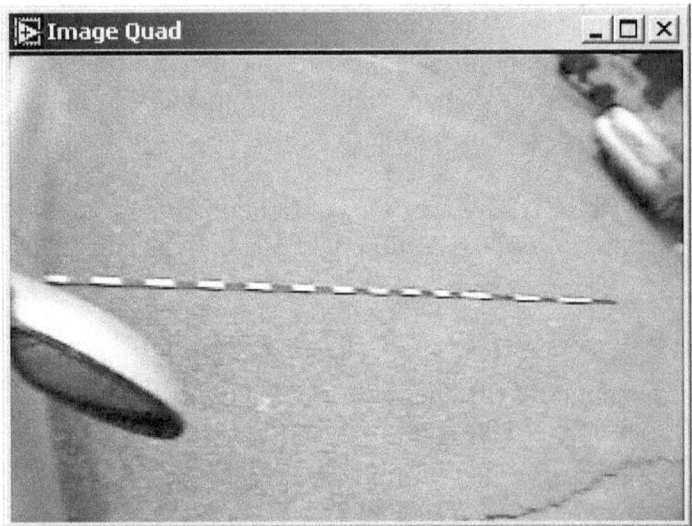

Figure 13. The top right quadrant appears in separate window.

4.2.3 User define cal points

The next step is to select the pixels corresponding to the endpoints of each segment in the meter sticks. If working with similar meter-stick configuration (for example when repeating a calibration or calibrating the side view cameras), turn on the "Use Default Cal Points" button (see Figure 10). In most cases, the button loads circles closely aligned with the segments. The user then moves the circles with the control panel arrows. The steps below describe defining the segments from scratch. Afterwards, you can rerun the vi and use the default cal points to experiment with adjusting the pixel coordinates. This is useful if you are not satisfied with the original calibration.

When the "User define cal points" is selected, the "Get Grid Point.vi" panel pops up (see Figure 14) and a text box appears with reminder information. For the top right quadrant the box states:
1. First grid point should be a bottom left.
2. Check Real World Grid Dim (m)

The first comment indicates that the first grid point should be at the left (bottom is for a two dimensional grid used for nonlinear calibration. The second comment reminds you that you must set "Real World Grid Dim (m)" to the actual distance between grid points. As stated above, the cal sticks vary for the side views. The user adjusts the distance value when the distance changes between each segment. Click OK to remove the text box and to start selecting the segment pixel points.

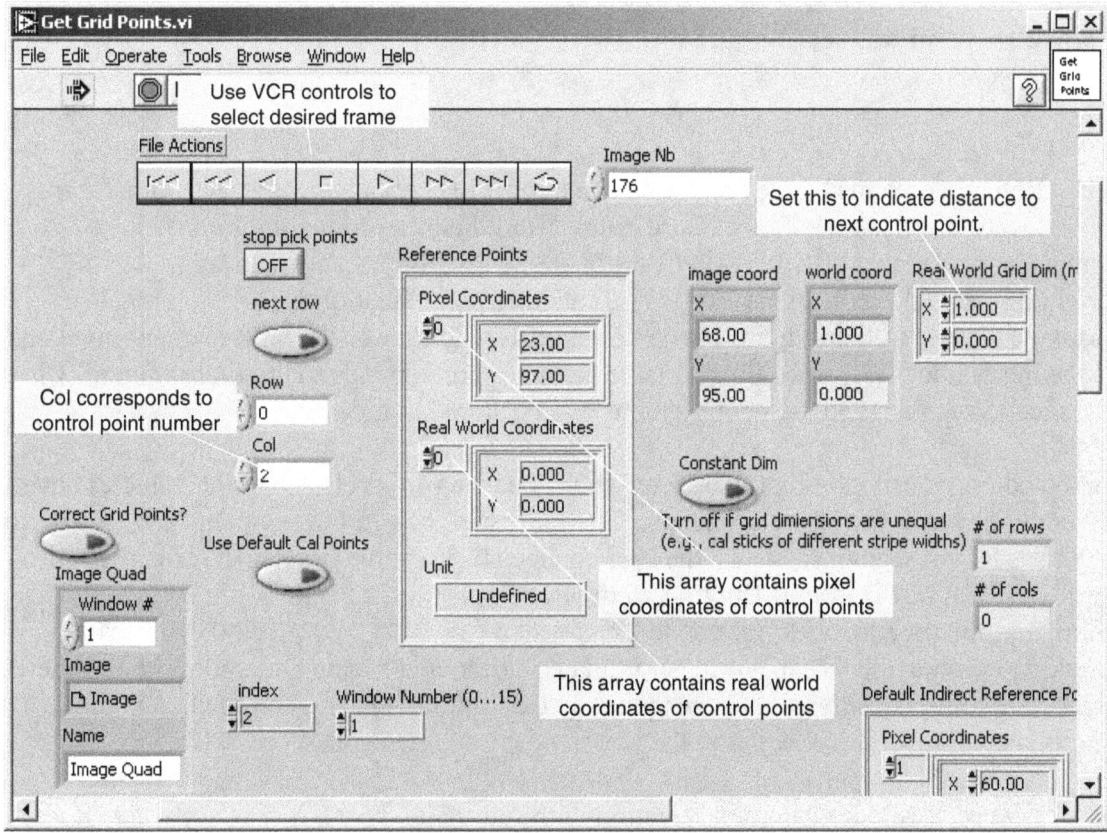

Figure 14. Control panel for selecting the cal stick pixels and strip dimensions.

To pick the first segment on the calibration stick, first make sure the "Image Quad" window is active by clicking on the blue bar above the image (see Figure 15). Then click on the pixel in the image that corresponds to the first point of the calibration sticks (in this case, the far left segment). The routine does not require perfect pixel selection. Select "Correct Grid Points" and the software will display a new panel with pixel adjustment directions. Grid adjustment is also available in a later step. Figure 15 shows the image with the first two calibrations segments marked.

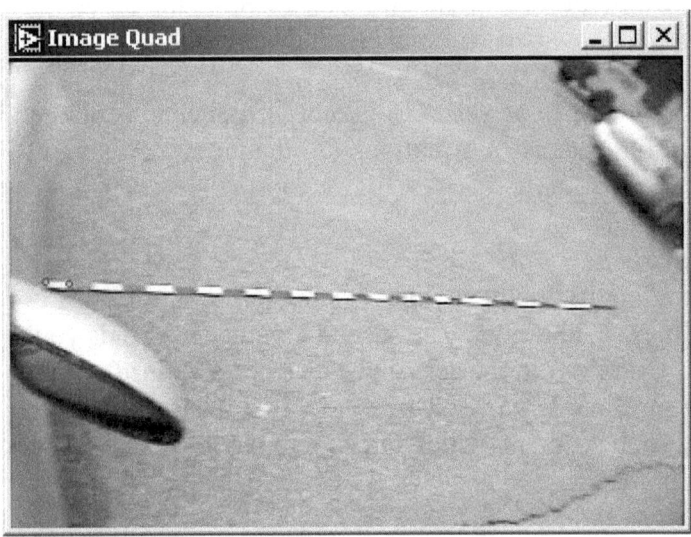

Figure 15. User has selected the first two segments.

The "Reference Points" table (center box in Figure 14) contains two arrays: one for the pixel locations ("labeled "Pixel Coordinates") and the other for calibration stick dimensions ("labeled "Real World Coordinates"). In this case, array location 0 is the origin of the calibration sticks (X=0, Y=0) and the pixel location is (X=23, Y=97). The display labeled "Col" in Figure 14 increments after the user clicks on a segment pixel and is the index into "Reference Points" table arrays (0 corresponds to the first segment, 1 is the second segment and the next segment will be stored at index 2).

As noted previously, the first two meter-sticks (starting from closest to the vehicle) have 10 cm wide black and white segments. The third stick has 25 cm segments and the fourth stick has 50 cm segments. Thus, the user changes the X value in "Real World Grid Dim (m)" (see Figure 14) from 0.1 m to 0.25 m after Col is 20 (i.e., 0 to 20 are 0.1 m apart). Similarly, the user changes X value to 0.5 m when Col is 25. To check the calibration process, examine the "Real World Coordinates" after clicking on the end of the last meter stick. Set the array index to 26 (the value of "Col" minus 1) and examine the X value. It should be 4.0 m, corresponding to the length of the 4 meter-sticks.

The operator concludes cal point selection with "stop pick points". The software automatically displays the "Move Reference Points.vi" control panel (Figure 16). Using this vi, the operator adjusts the pixel locations and/or changes the segment width. The control panel includes instructions in the upper left corner. The operator selects a single red circle with "Point #" and moves the circle with the sliders. Figure 17 shows control point number 3 highlighted (large green circle). The operator moves all the points simultaneously with "Move all points". The "stop" button concludes point correction and returns the operator to the "Nonlinear Calibration.vi" control panel.

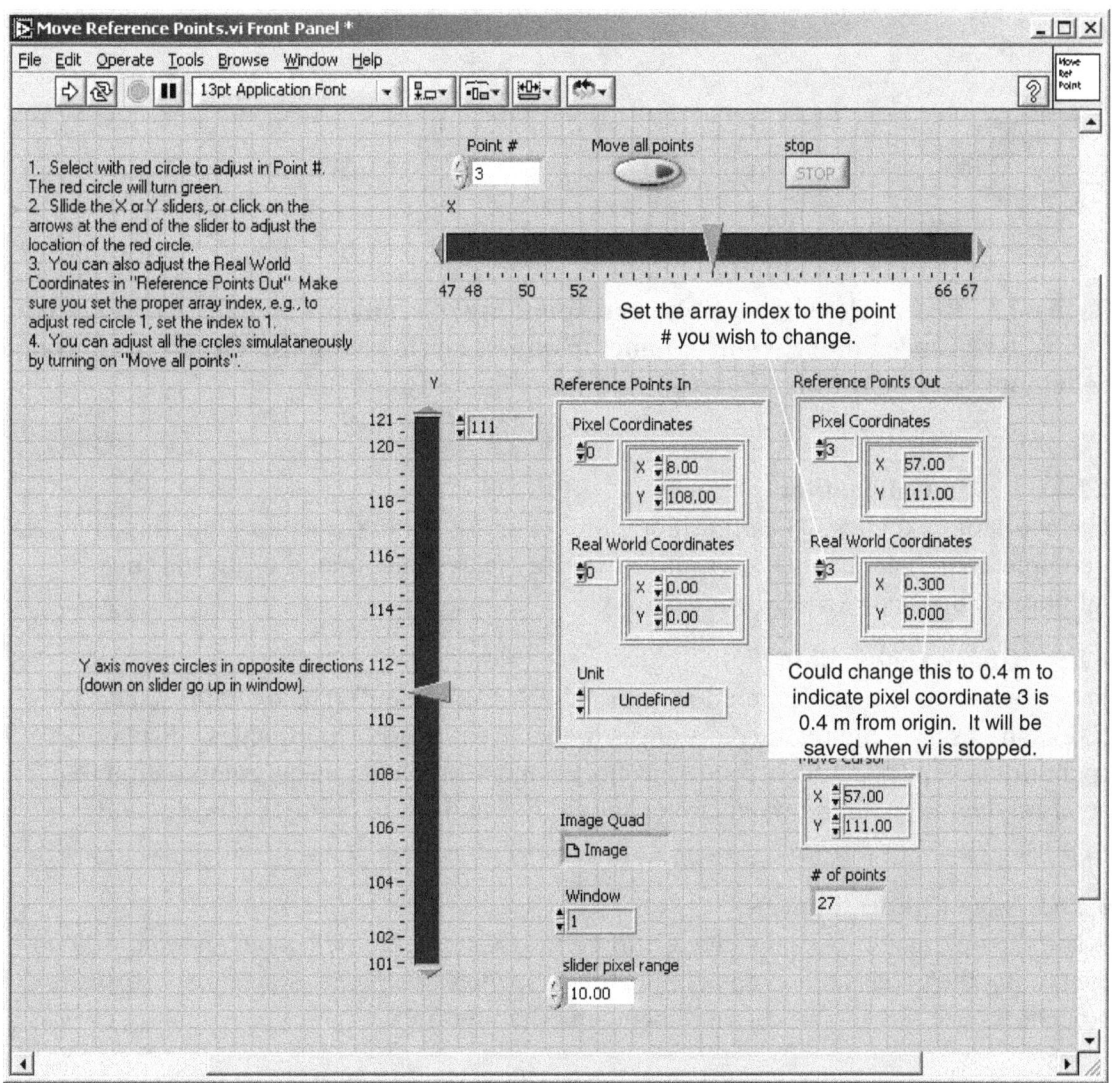

Figure 16. Control panel to move calibration points (red circles).

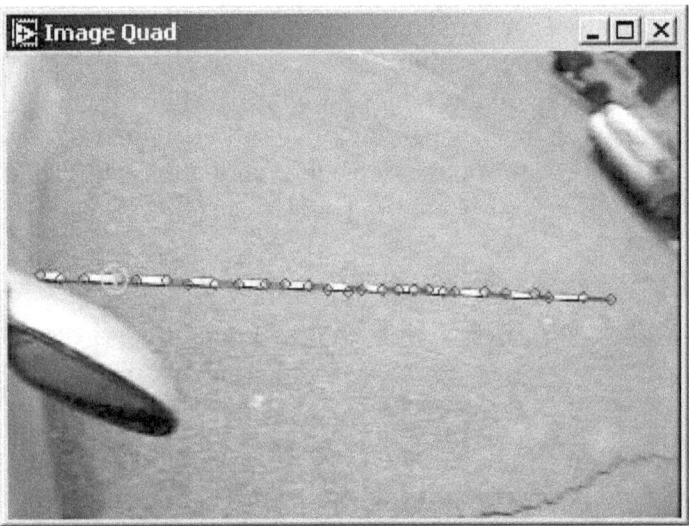

Figure 17. Red circles show pixels corresponding to segments on cal stick. Use the "Move Reference Points" control panel to adjust circle locations. Green circle is one being moved after setting "Point #" to 3 (see Figure 16).

4.2.4 Learn Calibration
The operator next clicks "Learn Calibration". For the nonlinear calibration, this invokes a complex nonlinear calibration routine that may take some time. For the cal stick calibration, the step executes immediately.

4.2.5 Check Calibration
The operator verifies the calibration with "Check calibration". The operator selects a pixel and observes the reported distances in the "Real World Coordinates" display. The operator also clicks pixels between the circles and verifies the linear interpolation (e.g., pixels at the mid point between circles should return the mid point distance in units fo length).

4.2.6 Save Calibration
When the operator clicks on "Save Calibration" the software writes out an ".ini" file containing the pixels and corresponding distance measurements. Subsequent software procedures use this file for measurements in other avi files.

A pop-up window asks if this calibration should become the default for the "Use Default Cal Points" described in section 4.2.3.

4.2.7 Return
Return ends the program. The operator re-runs the program for the other quadrants.

4.3 Considerations for Calibrating the Front View
As indicated earlier, the front view calibration varies slightly from the side calibration.

1. The front view calibration uses meter wide segments (see Figure 14). Therefore, the "Real World Grid Dim (m)" is 1.0 m.

2. The forward view control points often blend into the background. The operator uses the frame that best distinguishes the control point. To assist the calibration, users identify the points with a perpendicular stick during the recording (Section 3.2). Figure 18 shows the procedure and the corresponding pixel marked (red circle). The operator selects the video frame via the VCR controls (see "File Actions" in Figure 14).

Figure 18. A user pointing to the control points using a colored meter stick.

4.4 Creating an INI File for the Video Calibration

This file contains calibration information. The section above creates the file and the operator adds more information. The root file name is the same name as the calibration avi file (e.g., the INI file for "cal stick 03-14-05_c.avi" is "cal stick 03-14-05_c.ini"). The ini file contains subjects and values of the form:

```
[FRONT_ABC_DIM_METERS]
A=2.144
B=5.95
C=6.3245
[SIDE_GRID_LENGTH_METERS]
GRID_LENGTH=4
[GPS_ANTENNA_MEASUREMENTS]
TO_FRONT=2.8
TO_PASSENGER_SIDE=0.855
TO_GROUND=0.0
[VEHICLE]
NAME=Altima
WIDTH=1.71
WHEEL_BASE=2.82
FRONT_AXLE_TO_FRONT=0.96
```

REAR_AXLE_TO_REAR=0.00

The INI file data definitions are:

FRONT_ABC_DIM_METERS The	section ID for the three dimensional parameters (in meters) of the triangle defining the offset and orientation of the front cal sticks (see Figure 9 on page 7).
A	The distance (m) between the left end of the cal sticks and the intersection with dimension B
B	The distance (m) between the left rear wheel-axle and the closest point on the cal sticks. The tape measure should touch the front wheel in order to be parallel with the vehicle.
C	The distance (m) from the left rear wheel (at the axle) and the left end of the cal sticks.
SIDE_GRID_LENGTH_METERS The	section ID for the length of the side cal sticks. If four 1-meter sticks are used, then the length is 4 m.
GRID_LENGTH	The total distance (m) of the array of side cal sticks.
GPS_ANTENNA_MEASUREMENTS	The section ID for GPS antenna measurements (antenna position on vehicle).
TO_FRONT Distance	(m) from center of antenna to front of vehicle bumper.
TO_PASSENGER_SIDE Distance	(m) from center of antenna to outside of vehicle on passenger size.
TO_GROUND Distance	(m) from center of antenna to ground (in example above set to 0 since this is not used in any of the analyses).
VEHICLE	Section ID for vehicle parameters.
NAME	Vehicle name (generally the model name).
WIDTH	Vehicle width (m)
WHEEL_BASE	Distance between axles (m) measured from centers of each wheel.
FRONT_AXLE_TO_FRONT Distance	(m) between front axle and front bumper.
REAR_AXLE_TO_REAR Distance	(m) between rear axle and rear bumper (in example above set to 0 since this is not used in any of the analyses).

4.5 Calibration Verification

Between tests, or even during a test run, a camera may move. The video image contains fixed features such as hood, fender, and mirrors. The operator verifies the calibration legitimacy through the constant position of these features. The operator may correct misalignment during data collection by realigning the camera. The calibration process described in the previous sections is invalid for a misaligned camera. Unfortunately, due to the non-linear nature of the image, the operator cannot discern the proper calibration from the images taken by the misaligned camera.

The misalignment may be very slight. Figure 19 shows an example of video collected with the front-view camera misaligned. The lower quad picture is the calibration and the top quad picture is a nighttime test. Notice the left shift of the windshield sprayer nozzle on the hood. Lateral measurements using the front camera are therefore invalid and the evaluator may only use the side cameras. Fortunately, the operators detected the misalignment during data collection and corrected the camera before the subsequent run.

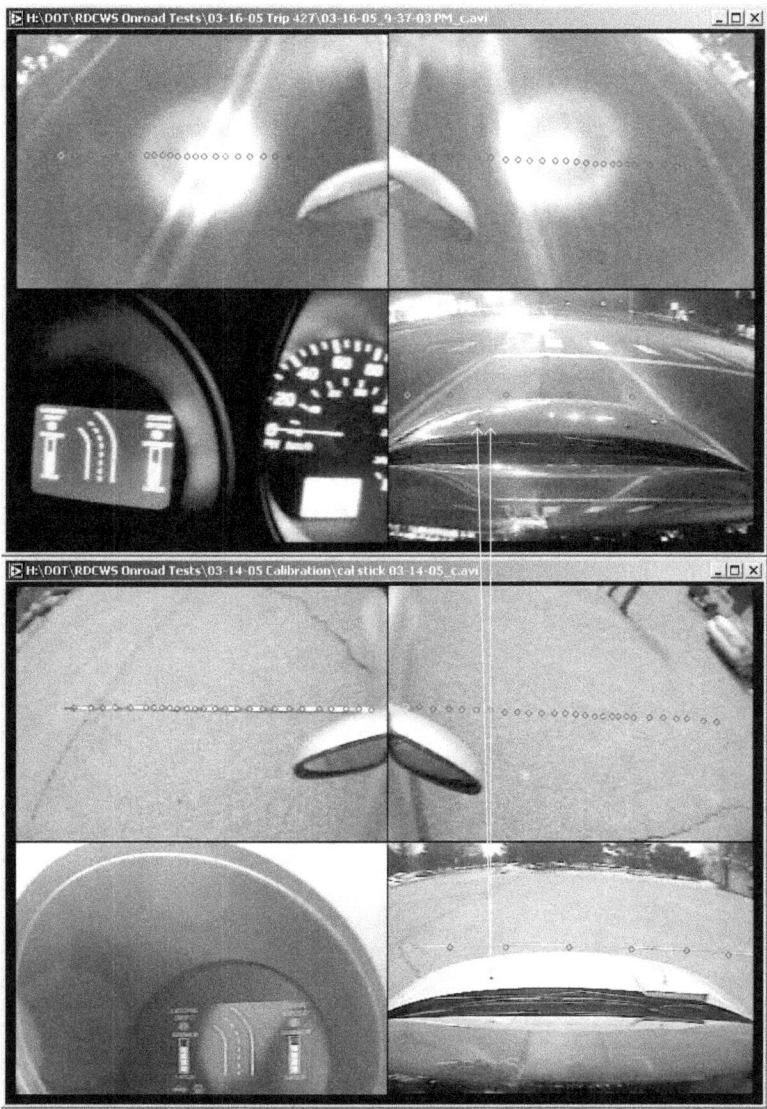

Figure 19. Example of camera misalignment.

5 Data Processing

Data processing prepares the data for analysis in LabView. The original data is on videotape. The first step transfers the video to a computer. NIST uses commercial software for the transfer (e.g., Adobe Premier). The evaluator examines the video and locates interesting segments (called clips). Clips capture a lateral drift with a warning, a missed warning (false negative), a specific run of a test identified in the RDCWS test procedures document, or some incident of interest. The commercial software locates the start point and stop point of the clip, encodes the video (DV format, e.g., Microsoft MPEG-4 Video Decompressor), and transfers the clip onto the disk.

The second processing step extracts the GPS data from the videotape's audio track. This step requires a VMS, a box that connects to the tape player, and Mediamapper software. The Mediamapper software extracts the GPS data and Easydiff software computes differential corrections. The software stores the GPS data in a file associated with the video clip file.

As implied above, data processing requires multiple files. Table 1 contains an example of the types of files created during data processing. The remainder of this chapter explains the data processing procedures.

Table 1 Data Processing File Examples

File Name	File Contents
10-09-03 Test 3.2..1.4.avi	An avi video clip containing video and audio from a run on 10-09-03 of Test 3.2.1.4 from the Visteon validation test plan.
10-09-03 Test 3.2..1.4_c.avi	A compressed version (see the _c at the end of the name) of the clip.
10-09-03 Test 3.2..1.4_c.ini	A text file containing information that is used during data processing. Important information contained in the file includes: • The video calibration file • The clip's location on the tape (time code of the clip; called "In Point" in Premier), which is used to synchronize the clip with GPS data
10-09-03 Test 3.2..1.4_c.vms	A text file containing GPS and tape time code information. Clips may not have a VMS file associated with it.

5.1 Time codes

The data processing uses various time codes (Table 2).

Name Source		Format
UTC	Universal Time Code. Time at the Greenwich Meridian	UTC from VMS file is "day.fraction of day"

		Extracted from GPS messages. Derived from GPS time (+ leap seconds) inserted by VMS onto tape.	This UTC is in seconds from start of day
	UTC Local Time	Local time derived from UTC – (time zone offset & daylight savings).	Seconds from start of day
	Time of Day	Entered into camcorder by user. This differs from UTC Local Time. The time is overlaid on tape but is not automatically transferred to computer upon capture (overlays do not transfer over firewire).	hh:mm:ss E.g., 3:40 PM
	Tape Time code	Essentially the tape counter. Defines a specific frame on the tape relative to the beginning of the tape. This time is needed to access video on a tape.	hh:mm:ss:## where hh = hour, mm = minute ss = second ## is the frame count
	Clip Time code	Defines a specific frame in a clip relative to the beginning of the clip. This time is used to access video in a clip.	hh:mm:ss:## where hh = hour, mm = minute ss = second ## is the frame count

Table 2 Time Codes

5.2 Video and Data Timing

The following discussion describes the procedure for determining the UTC time (and subsequently, the corresponding GPS coordinates) of a particular video frame in an avi file. The VMS inserts the UTC (from GPS data), the position (from GPS data), and the tape time code on the tape once per second. The tape time code is the time (from the start of the tape) when the VMS writes the GPS data. The tape time code does not align with the avi time (seconds from the start of the avi file) since the avi may be a clip extracted from the tape. The VMS data relates the video frame number to the tape time code and thus to the GPS time and position.

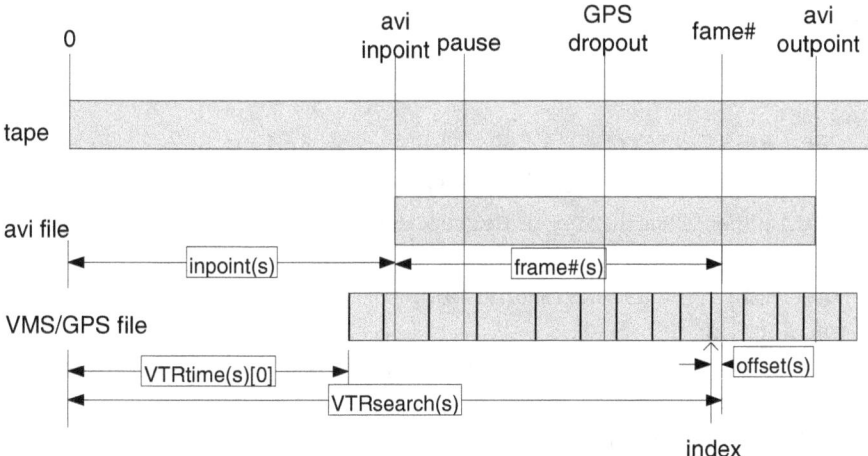

Figure 20. Timing diagram describing relationship between video tape, avi file and VMS/GPS file.

Figure 20 shows the relationship between the different data streams. Table 3 describes the terms in the figure.

Term Definition	
tape	magnetic media that holds video and VMS data (GPS time and position, and tape time code) at 1 Hz.
avi file	captured from some point on tape starting at avi inpoint and ending at avi outpoint. The points are stored in the INI file (discussed later) named after the avi file.
VMS/GPS file	captured from tape at some point before and after avi in and out points. Contains VTRtime, UTC time and GPS coordinates. These are stored into individual arrays called VTRtime and UTC.
VTRtime[0]	the first element in VTRtime is the starting time on the tape where the first VMS data is available. The format is hh;mm;ss;## where ## is the frame count within seconds.

Table 3 Time Reference Terms

The following steps determine the UTC time for an avi frame# and the index into the GPS position array corresponding to a video frame. Tape pauses or GPS dropouts do not affect the UTC time from these procedures. In the discussion, the string "(s)" indicates the data in seconds.

1. Convert frame# into frame#(s) based on frame rate.
2. Convert inpoint into inpoint(s).
3. Set VTRseach(s) = frame#(s) + inpoint(s). Want to search for this tape time code.
4. For each VTRtime[i=0..N]
    a. Calculate VTRtime(s) of VTRtime[i]
    b. Stop when VTRtime(s) is greater than VTRsearch(s) (i.e., went one record too far into tape)
5. Set index to i-1 (i.e., VTRsearch(s) falls between index and index+1)
6. Set offset(s)=VTRsearch(s) - VTRtime(s)[index]
7. Set Frame#(UTC)=UTC[index]+offset(s). This refines the UTC time based on the number of frames from the last UTC update.
8. Return index and Frame#(UTC).

5.3 Creating the avi file using Adobe Premier

Use a DV capture program to transfer video from a tape to a computer file. The following contains directions for capturing video and creating an avi file using Adobe Premier. Appendix A contains some Premier quirks and error conditions.

1. Open Premier and select the DV-NTSC Real-time Preview->Standard 48kHz setting.
2. Select File->Save as. This will bring up the save file window with an "Untitled.ppj" project file name. Create a folder that will have the same name as the tape name.
3. Create a bin. Within the project are bins. Typically, there is only one bin per project. Create a bin name according to the date of the videos were taken.
4. Select Edit->Preferences->Scratch Disks and Device Control. This selects where the clips are stored and the type of DV recorder used. The scratch disks should be the same folder as the project file. Set the "Same as Project File" to the project file. The

Device should be a DV Device Control 2.0. Check the options to select the correct camera (see Figure 21).

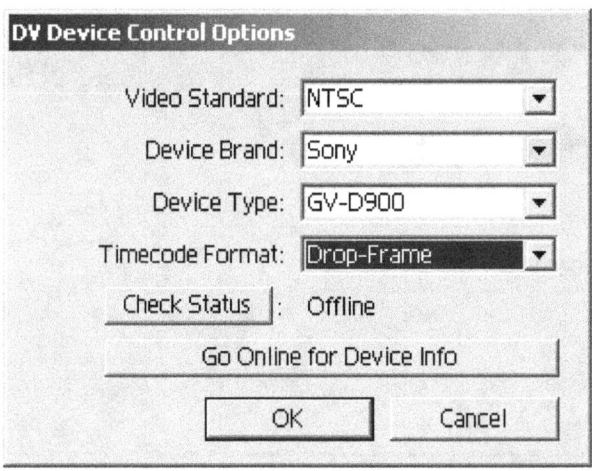

Figure 21. Settings for the Sony GV-D1000.

5. Select File->Capture->Batch Capture. This will bring up the Batch Capture window. Go through the tape and select start and stop points of various clips of interest. The clips will be stored in the batch capture window. When done, one batch run will capture all the clips. Save the batch window so that in the future the clips can be re-captured without having to search for the start/stop points. Immediately save this as "Batch List" in the project folder.
6. Select File->Capture->Movie Capture. This will bring up the Movie Capture window. Enter the name of the tape in the Reel Name field.
7. Use the controls to place the video at the start of the clip. Click on "Set In" to enter the start of the clip. Use the controls to place the video at the end of the clip and click on "Set Out". Then click on "Log In/Out". A window will come up asking for a name. Enter the name, such as "cal" or "test". Premier will automatically create new names for future clips by using the current name and adding an incrementing number. Repeat this process to log all the clips.
8. Go to the Batch Capture window and click on the record button (red button in lower tool bar). **Don't click on anything outside of Premier or the capture will abort.**
9. To view a clip, drag into the Monitor Source window and press the play button. The play performance may be poor if the video is also being output to the camcorder or DV deck (you will see the video displayed on the camcorder's monitor). To turn off video output to an external device, select Project->Project Settings->General, click on

"Playback Settings" and deselect "Playback on DV Camcorder /VCR".

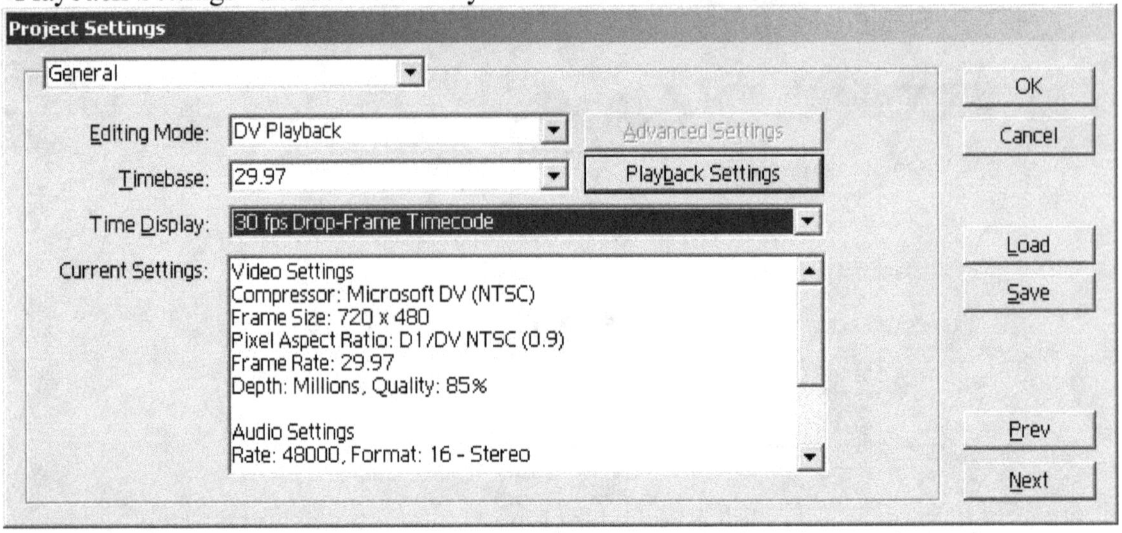

Figure 22. Project settings.

Insert the source video into the program. Drag the source window into the program window (both are located side by side in the Monitor window). The clip will also appear in the Timeline window.

For final output, it is desirable to remove the VMS data (a modem sound with bleeps every one second) from one of the stereo channels. This will make cabin sounds, driver instructions and audible warning much easier to hear. To eliminate the GPS audio channel, select an audio clip in the Timeline, right click and select audio options->duplicate right (or left). Alternatively, use the audio mixer (select Window->Audio Mixer) to only lower the VMS audio. First, select the middle button (looks like a pencil writing) above the Mute button to set the mixer to make permanent changes to the audio adjustments. Otherwise, all changes are lost when leaving the mixer window. Then adjust the balance (+98 works well, the VMS modem noise is barely audible). Then play through the entire clip (using the play button in the mixer) to write the new balance. The balance adjustment is lost if play is stopped before the end of the clip.

10. Compress the clip to save disk space.
    a. Drag video to start of timeline. Set the work area (yellow bar above time line) to enclose the clip.
    b. Select File->Export Timeline->Movie…
    c. Choose same name as original avi with appended "_c" to indicate the file is compressed.
    d. Select Settings. Go through each setting (using Next button) and set the same as shown in the following figures. Make sure to set the compression frame rate to 29.97 frame/s (not 30 frames/s – see Figure 24 and Appendix A).

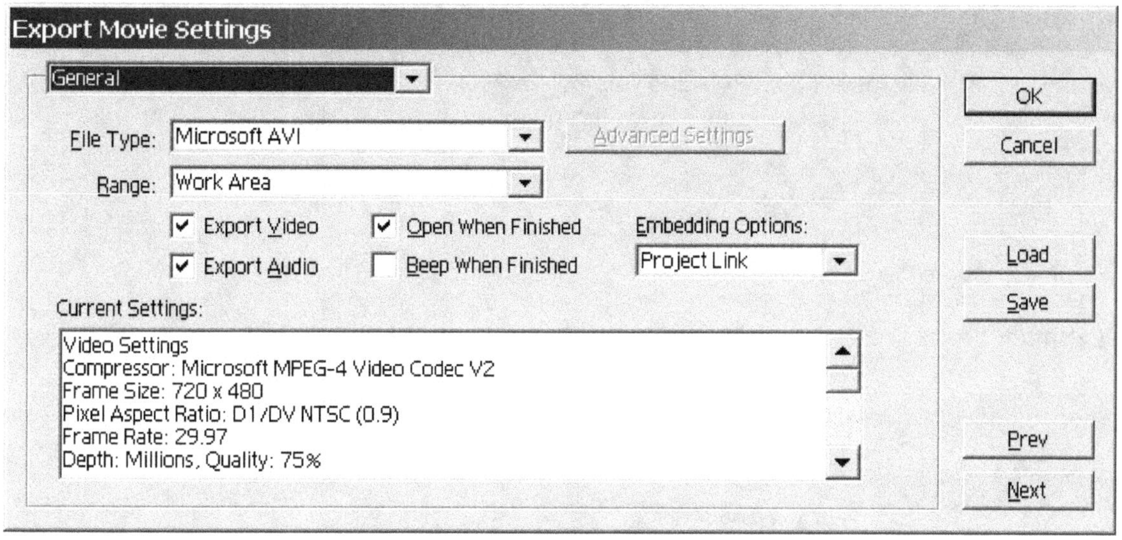

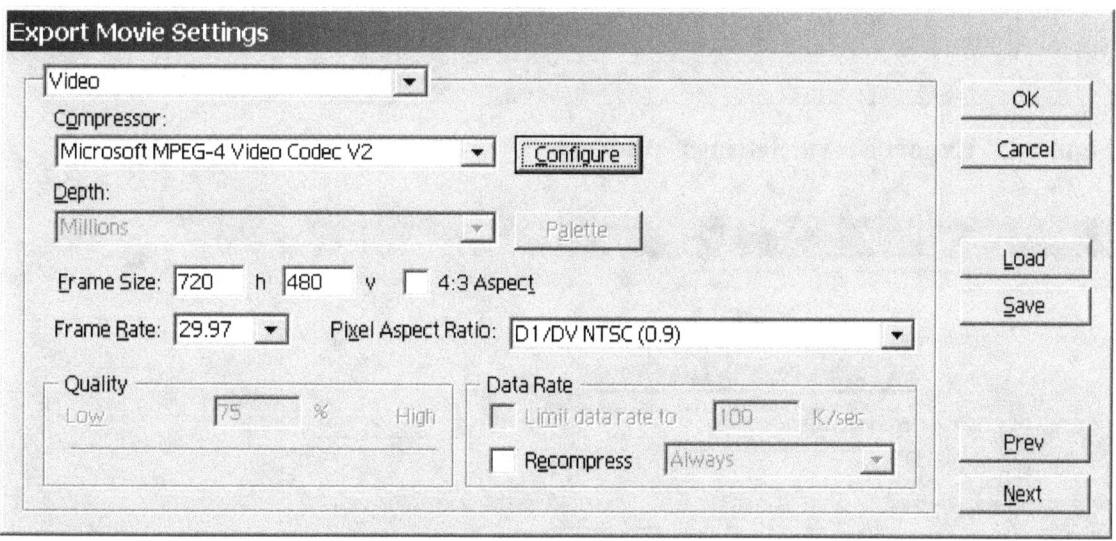

Figure 23. Export Movie Settings – General.

Figure 24. Export Movie Settings – Video.

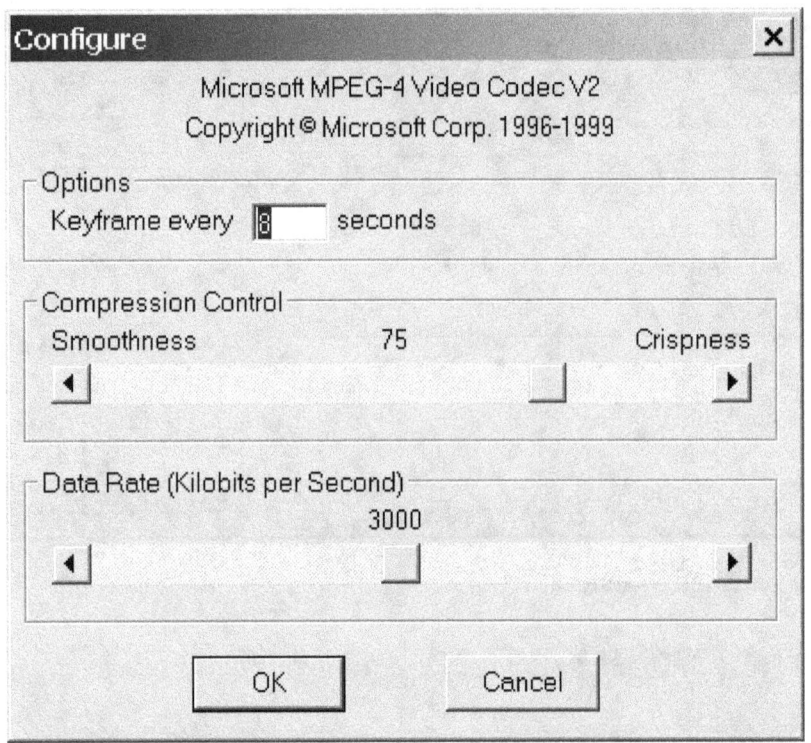

Figure 25. Export Movie Settings - Video->MPEG-4 Configure.

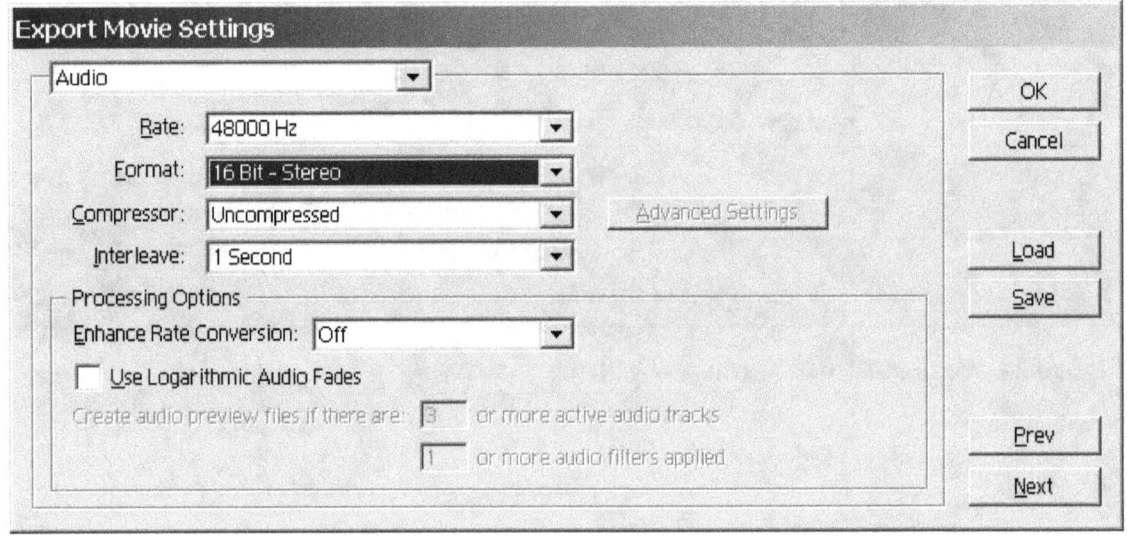

Figure 26. Export Movie Settings – Audio.

5.4 Creating the INI file

An INI file, created by the user (typically by modifying an existing INI file), stores data associated with the avi file and has the following form:

    [Video Data]
    In Point=00;02;07;00

Out Point=01;25;11;00
Local Start Time="02:59:08 PM"
Date=03-14-05
Frame Rate=29.97
Number of Frames=149371
Calibration Video File Name="cal stick 03-14-05_c.avi"

In Point	Start timecode (i.e., in point) of video clip. The timecode is relative from the start of the tape. It is the same as the "set in" in the Premier capture window or batch file.
Out Point	Stop timecode of video clip. Same as the "set out" time in Premier.
Local Start Time	Start time of video clip as displayed on tape counter. Since this is typically set manually, it is not very accurate. If available, GPS data is used for generating the local time once the VMS file has been created.
Date Date	
Frame Rate	Found from Premier capture program or by right clicking on the avi file in windows and clicking on properties (go to summary tab). Note that it is typically 29.97 frame per second and not 30 fps.
Number of Frames	Number of frames in avi file. Also available from avi properties.
Calibration Video File Name	Name of the avi file used to generate the calibration information for this avi file.

Table 4 INI File Definitions

5.5 Creating the VMS (GPS) file

The VMS file is a text file of Latitude, Longitude, speed and tape time code. Each line contains one data set and represents one-second intervals. See Appendix C for directions on how to extract the GPS data from the videotape.

6 AnalyzeData Main Control Panel

The evaluator uses the AnalyzeData program, written by NIST researchers, to analyze IMS data (Figure 27).

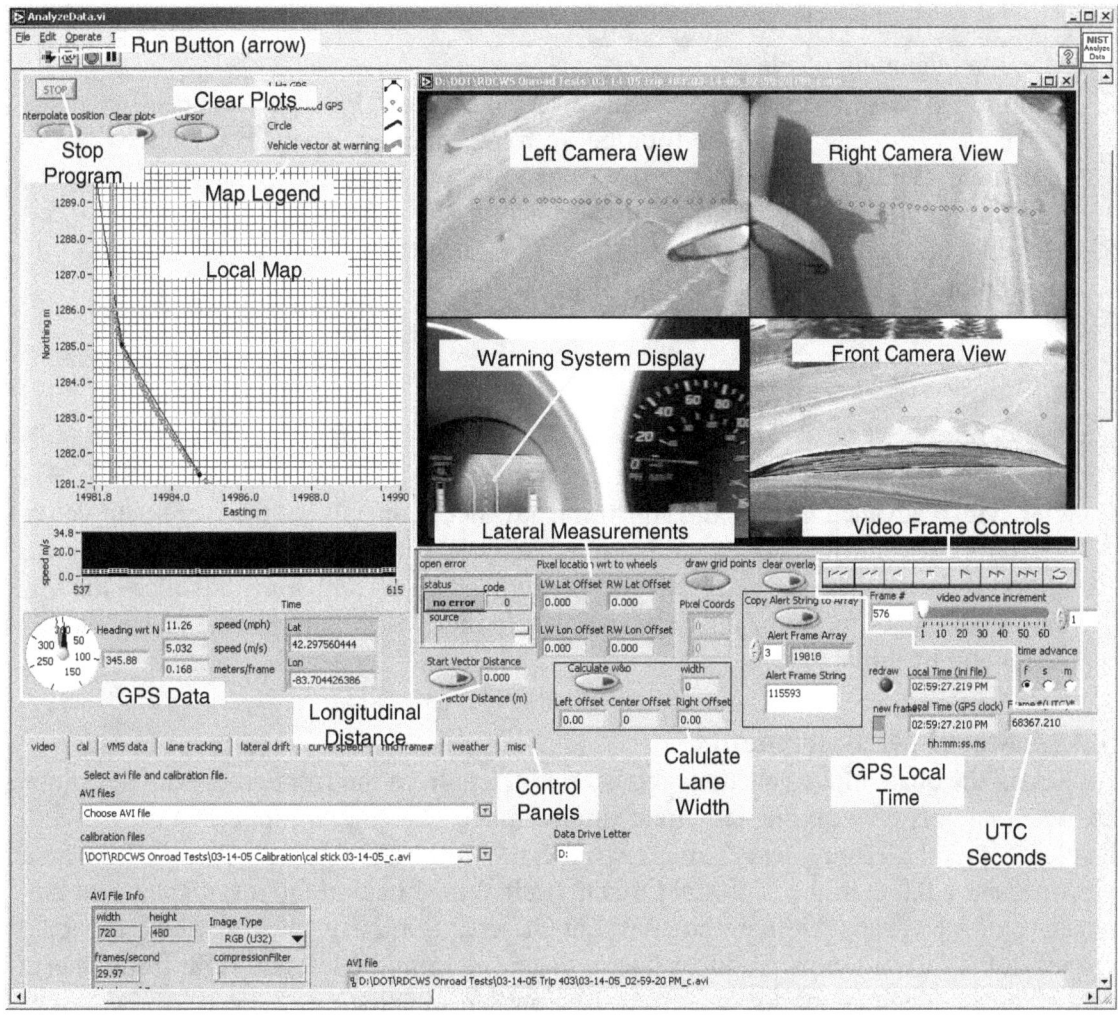

Figure 27. The AnalyzeData control panel.

The following are the controls, displays and procedures for the examination of data and the collection of measurements.

6.1 Starting and Stopping

The white arrow button in the upper left corner starts the program. The program begins with a file selection menu. The user selects the avi file to analyze and a quad image appears in the upper right.

The arrow button turns black during program execution. The operator stops the program with the red octagonal button next to the arrow button.

6.2 Video Frame Controls

The "Video Frame Controls" provide controls and displays for manipulating the avi file. The VCR control has eight buttons for changing the displayed frame (from left to right):

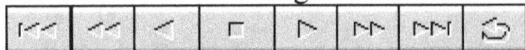

1. Move to beginning of file
2. Continuous backward play (may be slow depending on CPU and disk speed)
3. Move backward one increment (number of frames moved depends on "video advance increment" and "time advance".
4. Stop (when playing backwards or forwards)
5. Move forward one increment
6. Continuous forward play
7. Move to end of file
8. Continuous replay

Other video frame controls are:
- Frame # shows the number of the current frame (first frame is 0).
- The "video advance increment" and "time advance" controls set the increment value for the VCR controls. For example, to move by 10 s increments each time the play button is depressed, set "video advance increment" to 10 and "time advance" to "s". The other values for "time advance" are "f" (to increment by frames) and "m" (to increment by minutes).

6.3 Lateral Measurements

A click in the quad image generates the lateral offsets from the front wheels. The values are accurate only between the red circles in the image. The offsets appear in the panel labeled "Lateral Measurements". For example, the farthest right circle in the left camera view returns a 0.0 m in the "LW Lat Offset" (Left Wheel Lateral Offset) display and the negative of the vehicle width in the "RW Lat Offset". Similarly, clicking along the red circles in the forward view produces lateral distances from the left and right front wheels as well as the longitudinal distance from the front wheel axle (e.g., 3.13 m).

6.4 Lane Width and Vehicle Offset

The lane width and offset equations for the vehicle within the lane are (refer to Figure 28 for measurement descriptions):

```
LaneWidth(lw)= LeftOffset(l)+ VehicleWidth(vw) + RightOffset(r)
CenterOffset(co) = (RightOffset - LeftOffset)/2
```

$$lw = l + vw + r$$
$$lw/2 = l + vw/2 - co$$
$$co = lw/2 - l - vw/2$$
$$co = (l + vw + r)/2 - l - vw/2$$
$$co = r/2 - l/2 = (r)- l\,/2$$

NOTE: co is negative left of center

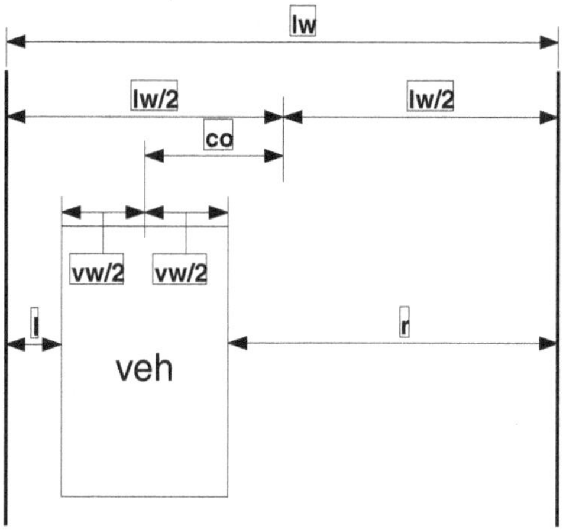

Figure 28. Lane width and vehicle center offset calculations and diagram.

The "Calculate Lane Width" controls (Figure 27) calculate the lane width and center offset. The operator uses the following procedure:
1. Click on inside of lane marker (aligned with red circles) on left side of vehicle (upper left quad. The value should appear in "Left Offset".
2. Click on inside of lane marker (aligned with red circles) on right side of vehicle (upper right quad. The value should appear in "Right Offset".
3. Click on "Calculate w&o"; the software writes the offset in "Center Offset" and the lane width in "width".

The software calculates the lane width with the vehicle width data from the video calibration INI file (see Section 4.4).

6.5 Longitudinal Measurements

The evaluator makes longitudinal measurements based on the GPS data. The Longitudinal Distance controls (Figure 27) compute a vector (straight-line) distance between any two video frames. For consistent measurements, the evaluator uses the same horizontal line in the two video frames, typically a red calibration circle. This measurement produces the vector distance and not the distance traveled on a curve (i.e., the arc length).

The longitudinal measurement procedure is:
1. Click on the "Start Vector Distance" button to start the distance measurement.
2. Advance the video frame. The straight-line distance is in "Vector Distance (m)"

6.6 GPS Data

The GPS data (position, velocity, heading, etc.) for each frame appears blow the local map (see Figure 28). The local map also shows the vehicle's GPS location. The units are in meters. The origin is the lowest Universal Transverse Mercator (UTM) coordinates of the GPS data from the current video file (see Appendix E for information on the UTM system).

The GPS data updates at 1 Hz. The video data updates at 29.97 Hz. On the map, the black circles connected by lines show the actual GPS positions while the red circles are the interpolated positions. The software interpolates the GPS positions with a second order polynomial. The local map maintains a 1:1 aspect ratio and automatically shows all the positions (i.e., the user does not need to scroll). The auto-zoom feature occasionally causes the map to look far-off. The user corrects the perception and clears the map with the "Clear plots" button.

The circle dial in the lower left shows the vehicle heading. The heading is the tangent to the second order polynomial interpolation of three GPS positions.

6.7 Control Panels

"Control Panels" is a tab switch across the bottom of the display. Each "Control Panels" tab displays a sub-panel containing controls and indicators for specialized analyses. The next several sections explain various analyses performed using sub-panel displays and controls.

7 Locating Video Based on GPS Time

The following section describes a procedure for locating video frames specified by GPS time. The software uses this procedure extensively when analyzing warning system data collected by the UMTRI's DAS. The DAS records the time of each warning issued by the Road Departure Crash Warning System. For on-road test analysis, Volpe queries the DAS database and generates a spreadsheet containing GpsTime and CspGpsTime (from curve speed warning system) of each warning. The procedure below takes each GPS time and locates the corresponding IMS video frame. The warning appears in the dash video quadrant (lower left) at some time nearby.

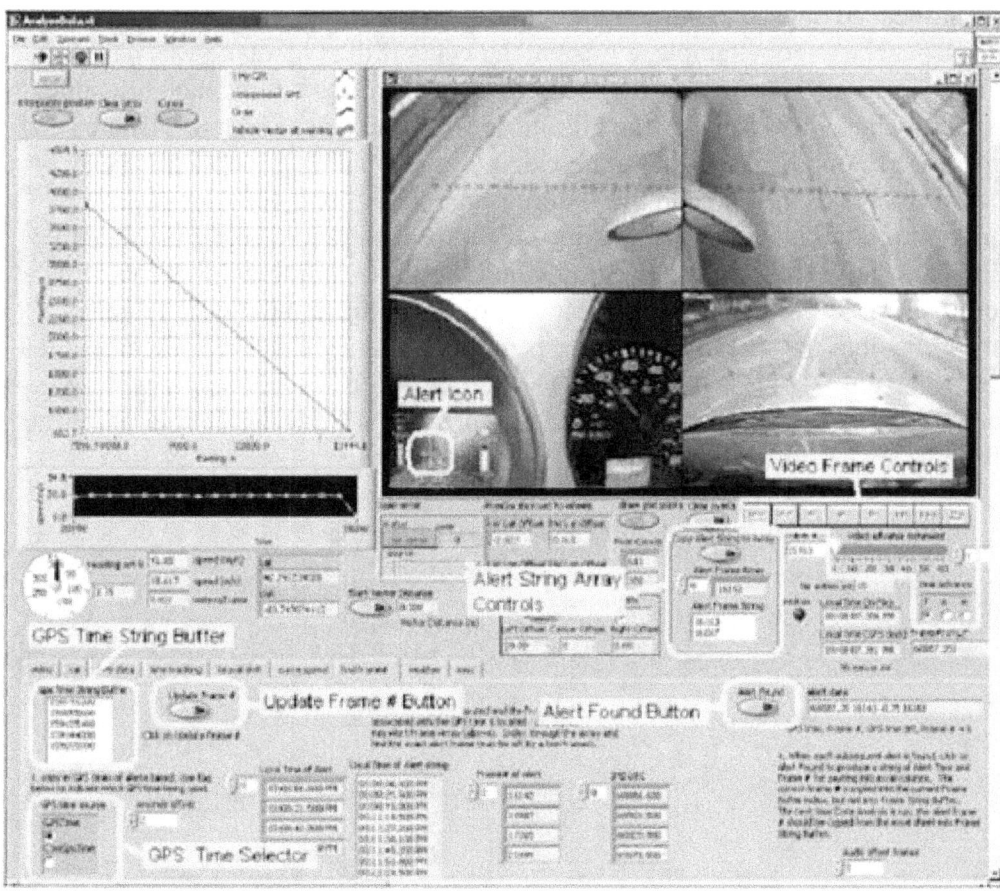

Figure 29. The "find frame #" Screen.

The frame to time correlation function is part of the AnalyzeData program. The evaluator uses the "find frame #" tab in AnalyzeData.vi (see Figure 29). The instructions appear on the tab and below.

1. Copy in GPS times "GPS Time String Buffer" (if values in buffer already, double click to highlight values and then paste to overwrite). The GPS times are a column in the Volpe spreadsheet.

2. Indicate the format of the GPS time with the "GPS time source" radio button. The choices are GPSTime and CswGpsTime from the Volpe spreadsheet header row.
3. Click on the "Update Frame #" button.
4. The local time and the frame# associated with the GPS time is computed and copied into the "Alert Frame Array".
5. Use the up/down array control to select an element of the "Alert Frame Array". The video frame automatically displays. Double check that the alert frame array number appears in "Frame #". Sometimes the update will be slow due to delays in reading the video file.
6. If the alert icon is visible, rewind the video to the first instance of the icon. If the alert icon is not visible, advance the video to the first instance of the icon. The "first instance" is the first video frame that includes the icon image. If the icon appears faded or hashed, advance one more frame. The estimate should be close, but there may be mismatches. For example, a mismatch occurs if the videotape was paused during the test drive. During a pause, the GPS time advances but the tape time does not advance. Thus, there is a jump in GPS time when the tape (i.e., avi) resumes.
7. Click the "Alert Found" button when the video frame containing the alert appears. The software changes the value in the "Alert Frame Array" and generates a string containing the GPS Time, the Frame #, the GPS time difference (between the Volpe GPSTime and the video frame's GPS time), and the value of frame# plus five. (The "5 +" corrects for a 0.17 s delay between the audio warning and the visual display.)
8. Paste the string into the spreadsheet (there is no need to copy since it is already in the clipboard).
9. Increment the "Alert Frame Array" index and repeat steps 6 through 9 until all of the Alert frame # are identified.

The "Alert Frame Array" is now ready for Lateral Drift Analysis.

8 Lateral Drift Analysis

The following lateral drift analysis uses the on-road characterization designed jointly by Volpe and NIST [2]. A Lateral Drift Analysis of a warning produces 23 values in six areas. The evaluator copies these values into a spreadsheet for a final report. Of the 23 values, two values deal with the time of the warning. Two more specify the nature of the warning. Three values deal with atmospheric conditions. Eight values describe the roadway's characteristics. Five values identify the vehicle's lane position and motion at the time of the warnings. The remaining three values describe the general operating conditions and the accuracy of the warning.

8.1 Overview

The following section gives a brief overview of the meaning of the values collected during a lateral drift analysis. See [2] for more details on these values. Section 8.2 presents a support tool for making measurements and generating entries.

8.1.1 Warning Time

The Lateral Drift Analysis records two warning times. The analysis presents both times by their video frame number. The first measure (Alert Frame #) is the frame when the alert icon is first fully visible on the video image. The second measure (Frame #) is the frame when the audible alert occurs. Evaluators identify the Alert Frame # through careful stepping through the video file (see Section 7). The Frame # is derived from the Alert Frame # and is used in subsequent evaluations.

8.1.2 Weather Environment

The Lateral Drift Analysis records three values describing the atmospheric conditions. From the video image, the evaluator determines if the alert occurred during the day or night, whether the road surface was wet or dry, and whether the driver was using the windshield wipers (i.e., was it raining). The evaluators observe sunlight as described in Section 8.4.

The day/night value is determined by comparison between the local time and the time of civil twilight. The local time of the warning comes from the GPS data. Civil twilight, which is the time after sunset when the sky becomes dark, comes from the United States Naval Observatory (http://aa.usno.navy.mil/data/docs/RS_OneDay.html). If the alert time is between "Begin civil twilight" in the morning and the "End civil twilight" at night, the evaluator records a "0" in the analysis. Otherwise the warning occurred at night and the evaluator records a "1".

Road surface wetness is a subjective evaluation. The evaluator views the road surface through the forward and side images and estimates if the road surface is wet. The evaluator records a "1" when the road is wet and a "0" when the road appears dry.

Wiper use is very difficult to determine from the available data. In some cases, the vehicle's hood may reflect the wiper motion in the forward view image. This is an unreliable, but only available, source for the data. The evaluator enters a "1" when wiper motion is detected and a "0" otherwise.

8.1.3 Warning Type

The Lateral Drift Analysis contains two values to identify the alert. The warning system generates imminent and cautionary alerts in both the right and left directions. The alerts, captured by the dash camera (see lower left quadrant of video), appear as icons on the warning system display. The evaluator enters a "0" for a cautionary alert and a "1" for an imminent alert. For departure direction, enter a "-1" when the alert is to the left, and a "1" when the alert is to the right.

8.1.4 Road Type

The Lateral Drift Analysis produces eight values describing the road. The analysis describes the type, the color, and the condition of the pavement markings in the alert direction. Two values record the direction of a curve and the vehicle's position in a curve. One value records whether the road is a freeway. The remaining two values describe the Available Maneuver Room.

8.1.4.1 Marker Condition

The analysis uses a three-tier evaluation for marker condition. The evaluator records a "2" if the quality of the markings is good, a "1" if the quality is fair, and a "0" if the quality is poor. A good quality mark is clear, crisp and bright. If the marks are faded, have irregular edges, or contain tar stripes, the evaluator records them as fair. We consider the marking to be of poor quality if it is missing, broken, obscured by debris, or is partly faded beyond recognition. When the quality of the marking varies, the evaluator records the markings deemed more significant to the alert.

8.1.4.2 Marker Color

Pavements marking are either yellow or white. The evaluator records a "1" if the markings are yellow and a "0" if they are white. If the color is indistinguishable, the evaluator records the color typically found at the subject location based on the marker type.

8.1.4.3 Marker Type

The analysis recognizes five marker types. The evaluator records a "0" when the approach toward a single solid line triggers the alert. The evaluator indicates a single dashed line with a "1". A double solid line garners a "2". The evaluator records a "3" when the lane marker is solid-dashed indicating the test vehicle can legally pass. The evaluator records a "4" when there is a solid-dashed combination where the test vehicle cannot legally pass.

8.1.4.4 Curve Direction

A three-tier value records the direction of a curve. A "1" indicates a right curve. A "-1" indicates the road is curving to the left. A "0" indicates the road is straight. The evaluator determines the direction of the curve from the map. The approach commonly used was to draw the curvature of the road by moving the vehicle from 1 s before to 1 s after the warning using one-second increments (use VCR controls and set increment to 1 s).

8.1.4.5 Curve Entry/Exit
The evaluator records the vehicle's relation to a curve at the time of the warning. When the alert occurs as the test vehicle enters the curve, the evaluator records a "1". The evaluator records a "3" if the test vehicle is exiting the curve. A "2" indicates the test vehicle is in the middle of a curve. The evaluator records a "0" when the alert occurred on a straight road section. Use the local map to determine the location of the vehicle relative to the curve.

8.1.4.6 Freeway Indication
A freeway is a road with a legal speed limit at or above 80 km/hr (50 mph) and a separation between the lanes going in opposite travel directions. The lane separation must have either a substantial vegetated area or a fixed barrier. The evaluator judges the road type from the video images and records a "1" for a freeway alert or a "0" for a non-freeway alert.

8.1.4.7 AMR
The Available Maneuver Room (AMR) is the distance between the inside edge of the lane marker and the nearest continuous obstacle or an opposing traffic lane. Continuous obstacles appear directly to the side of the vehicle. The vehicle approaches the obstacle at a slow rate. The warning system must be able to detect four types of continuous obstacles:
1. Concrete Traffic Barrier (i.e., Jersey Barrier)
2. Guard rail
3. Fence
4. Wall

The evaluator uses the calibrated video image to make lateral measurements of the AMR measurement (see section 6.3). The outside of the AMR is the edge of pavement, the bottom of an obstacle or, for an opposing traffic lane, the middle of the double lane markers or outside edge of the single lane marker.

8.1.4.8 AMR Type
The lateral drift analysis provides extra information for a left warning. The evaluator uses the forward and left side images to assess the area adjacent to the travel lane. The evaluator records a "0" if the adjacent area is the roadway shoulder. A "1" indicates the adjacent lane is a travel lane in the same direction as the test vehicle. A "2" indicates the drift is towards a travel lane in the opposite direction as the test vehicle. The evaluator records a "3" if the adjacent lane is shared, such as for passing. A "4" indicates the adjacent lane is not a travel lane, but is used for turning.

8.1.5 Vehicle's Lane Information
The Lateral Drift Analysis records up to five values describing the vehicle's relationship to the lane and roadside objects. Two readings specify the vehicle position within the lane. Another reading is the lateral velocity of the vehicle. When the vehicle's radar

units identify a discrete obstacle, two more readings identify the vehicle's position from the obstacle.

8.1.5.1 Lane Position

The Lateral Drift Analysis records the vehicle's lane position as two values: the distance from the left lane marker to the left wheel and the distance from the right lane marker to the right wheel. The evaluator uses the calibrated image to measure the distance to the lane marker from the wheel. The analysis software combines the left and right lane marker distances with knowledge of the vehicle's width to determine the lane width and the distance from the vehicle center to the lane center (see section 6.4), however, the width is not entered into the analysis spreadsheet (Volpe chose to calculate this value).

The lane width varies little over short sections of roadways. Therefore, once the analysis software has the lane width, only one lane marker is required to determine the distance from the center of the lane to the vehicle. The analysis software uses these facts to determine the distance to the lane edge when the lane marker is under the vehicle.

8.1.5.2 Lateral Velocity

The lateral velocity is the range rate, or average change in distance over time, between the vehicle and the lane edge. The video provides an excellent time stamp for lateral velocity calculations. The evaluator determines the lateral velocity from the lane offset from two video images. The evaluator selects images such that the lateral motion is easily discernable and consistent up to the alert frame.

8.1.5.3 Discrete Obstacles

For these tests, discrete obstacles are objects that appear on the shoulder ahead of the vehicle. The warning system must be able to detect five types of discrete obstacles:
1. Parked vehicle
2. Bridge structure
3. Sign Post
4. Pedestrian/Cyclist
5. Animal

The Road Departure Crash Warning System has a set of radars that alert the driver to approaching obstacles. UMTRI's analysis of the DAS data includes the source (spreadsheet column header AmrCriticalSource) of the alert and the approximate range (spreadsheet column header AmrCriticalBin). A source value of 2 indicates the radar generated the alert. The bin refers to half-second look-ahead times. A bin value of 1 (0 s to 0.5 s) indicates the object was detected in the side radar. A bin value of 2 (0.5 s to 1.0 s) and above indicates the object was detected in the forward radar. The evaluation includes the longitudinal and lateral position of the object.

The procedure for determining the longitudinal distance to an obstacle (requires GPS or speedometer data) is described in Section 6.5. The program enters the longitudinal distance in the XO column of the analysis spreadsheet. The obstacle's lateral offset is determined using the technique for measuring the AMR discussed above. The lateral distance is entered in the YO column of the analysis spreadsheet.

8.1.6 Warning Classifications

The Lateral Drift Analysis calls for three types of situation evaluations. The first describes the scenario of the alert. The second identifies any particularly challenging aspect of the environment or activity at the point of the alert. The third evaluation rates the accuracy of the alert.

8.1.6.1 Scenario Identification

The evaluator views the situation in the video and determines which scenario best describes the situation at the time of the alert. Table 5 lists the scenario identifications. The evaluator records the three-digit reference value based on the appropriate description.

Ref	Description
2.4.a	Departure on rural roads with narrow shoulder ≤ 1 m
2.4.b	Departure on roads with medium shoulder 1–2m
2.4.c	Departure on highway with wide shoulder ≥ 2 m
2.4.d	Drift left towards solid lane boundary
2.4.e	Drift left towards striped lane boundary
2.4.f	Drift right towards striped lane boundary
2.6.a	Drift left towards vehicle traveling in adjacent lane
2.6.b	Drift right towards vehicle traveling in adjacent lane
2.6.c	Drift towards guard rails in close proximity (≤ 1 m) to lane edge
2.6.d	Drift towards jersey barriers in close proximity (≤ 1 m) to lane edge
2.6.e	Drift towards traffic barrels in close proximity (≤ 1 m) to lane edge
2.7.a	Drift towards vehicle parked on shoulder
2.7.b	Drift towards bridge abutment
2.7.c	Drift towards telephone pole
2.7.d	Drift towards obstacle in curve

Table 5 Scenario Identifications

The scenarios cover three basic categories: lane departures, drift towards close objects, and drift towards other objects. Some departure alerts may include more than one scenario. When there are conflicts, the evaluator selects the most specific scenario. Therefore, the "drift towards close objects" (2.6.a through 2.6.e) has the highest priority. "Drift towards other objects" (2.7.a through 2.7.d) have the second highest priority. Finally, lane departure scenarios (2.4.a through 2.4.f) have the least priority.

8.1.6.2 Challenging Environment

The second Lateral Drift Analysis situation evaluation describes any particularly challenging aspect of the environment or activity at the point of the alert. Table 6 shows the codes for identifying challenging environments. The evaluator identifies and records the challenging environment.

Key	Environment
-1	none
0	glare due to low sun angle
1	shadows cast from trees
2	nighttime streetlight lighting
3	glare due to oncoming headlamps
4	roadside bushes
5	roadside sign
6	roadside guardrail
7	adjacent vehicle
8	roadside mailbox
9	roadside barrier (leading edge)

Table 6 Challenging Environment Codes

8.1.6.3 Alert Ratings

The third Lateral Drift Analysis situation evaluation records the accuracy of the alerts generated by the warning system. The alert may be either positive or negative and may be either true or false. Positive indicates the warning system generated an alert. True indicates the alert, or lack of alert, reflected the activity on the road.

The evaluator assigns a "1" for a true positive. A true positive rating indicates the vehicle either departed its travel lane or made a substantial move to depart its travel lane and the warning system sounded on alert. The evaluator assigns a "-1" for a false positive indicating the driver received an alert but the vehicle was not departing the lane.

The test vehicle has a button that allows the driver to record a particular time as a point of interest. The points of interest generally do not coincide with the alerts from the warning system. Thus, most points of interest are negative alerts. When the system does not issue an alert when the test vehicle departs the lane, the evaluator records a "-2" for a false negative. The driver may mark a point in the run when the vehicle passes a potentially difficult situation and responds correctly. A true negative generates an analysis and assists subsequent evaluations by including specific situations in the experiment statistics. The evaluator leaves the Alert Ratings entry blank for a true negative.

On rare occasions, the evidence from the video file is not sufficient to evaluate the accuracy on an alert. The evaluator records a "0" to indicate further study is required. The "0" entry keeps the event from biasing subsequent statistics.

8.2 Positive Alert Measurement Procedure

The procedure below steps the operator through the evaluation of a lateral drift event. Figure 30 is the operator's window for lateral drift evaluation. Refer to Figure 30 for component identification. Identify the starting video frame number when the alert occurs and enter the frame numbers into the Alert Frame String and Alert Frame Array before using this procedure.

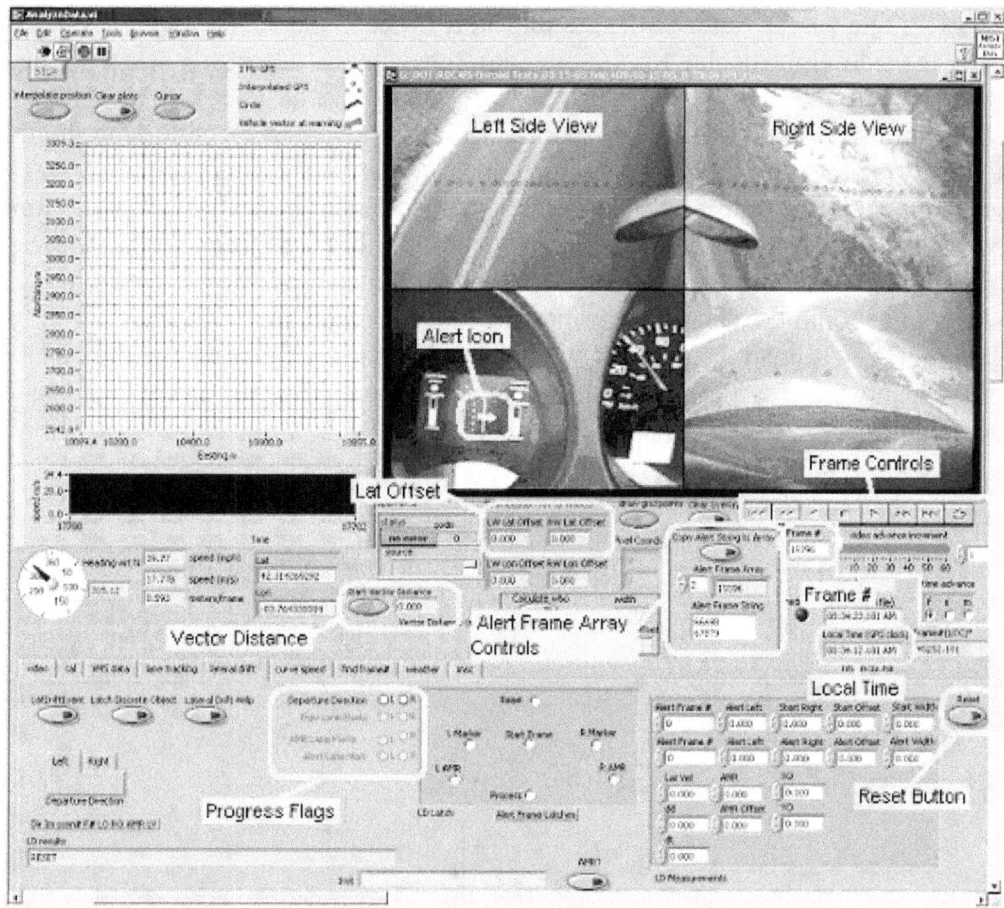

Figure 30. Lateral Drift Window, Method B.

1. Hit the "Reset" button (far right on "lateral drift" control panel) to clear all data from a previous analysis.
2. Select the next Alert Frame by either entering the number in the "Alert Frame Array" index or by clicking on the up triangle in the button next to the index (see Alert Frame Area in Figure 30).
3. Choose the Departure direction by clicking anywhere in the side-view display in the direction of the departure. The Departure Direction Flag will indicate your selection. If you select incorrectly, use the "Reset" button and start again.
4. Select the appropriate "Curve Info", "Alert Type", "Road Type", "Scenario", "Environment", "Marker", "Marker Quality", "Wiper", "Day/Night", and "Alert Rating" in the "drift filler" panel (see Figure 31). Reselect the values if cleared on Reset. Click on "keep values" to prevent the "Reset" button from clearing the "drift filler" values. This is helpful because it eliminates the need to continually re-select each value if the same scenario occurs several times. Note that the computer will beep if one or more of the evaluations are not complete.
5. Back up the avi display about 20 frames. Continue backing until both the right and left lane markers are visible.
6. Using the side-view displays, select the left and right lane markers. The "Prev Lane Marks" flag will indicate your progress.
7. Using the side-view displays, select the AMR marker in the departure direction.

8. The avi frame will advance to the audible alert frame.
9. Using the side-view display for the direction opposite the departure, select the lane marker.
10. If the warning stems from a discrete obstacle ahead (e.g., parked car on the shoulder), then use the procedure described in Section 6.5 to perform a longitudinal measurement to the object. The start of the measurement is the frame# of the warning and the end of the measurement occurs when the calibration circles in the side view align as close as possible to the object. Click on "Latch Discrete Object" to copy the measurement onto the clipboard for pasting into the result spreadsheet.
11. The program will load the operating system's clipboard with the lateral drift data. Select the next line on the spreadsheet containing the DAS warning data and paste lateral drift data.

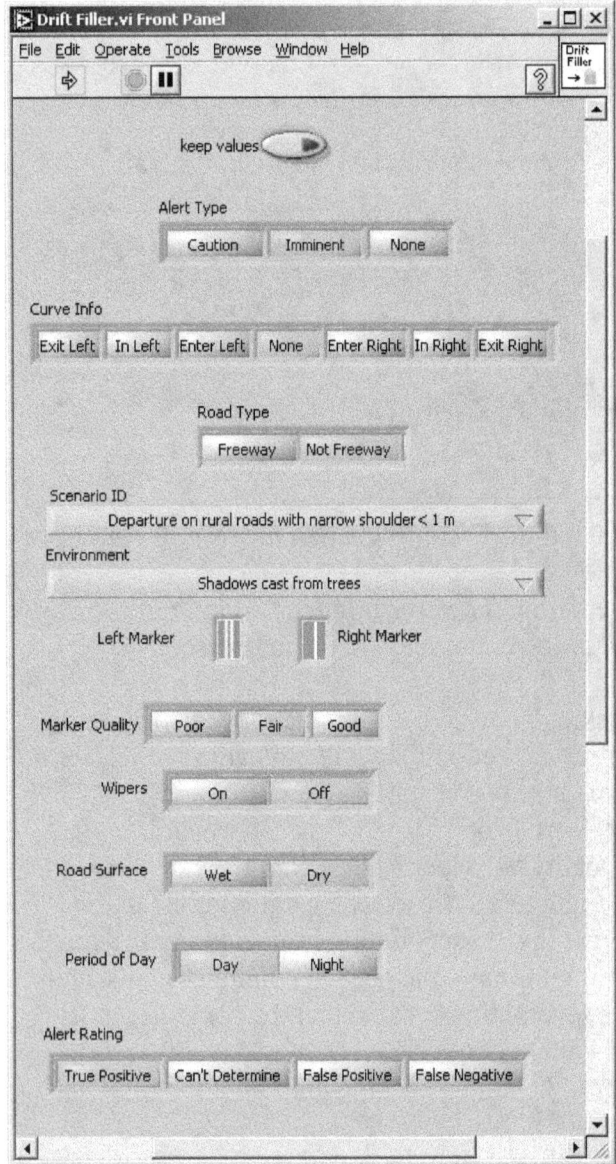

Figure 31. Drift Filler Window.

8.3 Negative Alert Measurement Procedure

The software coordinates the Positive Alert Measurement Procedure through the Alert Frame array and index. Negative alerts do not have an accurate Alert Frame. Negative alerts have only an approximate time of occurrence. The measurement procedure for negative alerts is similar to the positive alert measurement procedure with an initial step to determine the alert frame.

The first step in a negative alert measurement is to determine the alert time. Since the driver signaled the negative alert with the dashboard button after the alert system failed to warn of an event, the negative alert frame is before the recorded time. The evaluator begins with the recorded time. The evaluator adjusts the Frame # until the "Local Time (GPS clock)" matches the recorded time of the alert. The evaluator rewinds the video to the frame where the drift or departure stops. When the lane departure is a lane change, the desired frame is when the lane marker is half way across the hood in the forward camera view. For other departures, the desired frame is where the drift stops (since the evaluation is using a rewind, it will appear to be when the drift starts back to the lane). The evaluator records the desired frame at the end of the Alert Frame String, clicks on "Copy Alert String to Array", and selects the last "Alert String Array" index. Afterwards the evaluation follows the Positive Alert Measurement procedure.

8.4 Atmospheric/Cloud Indication

Atmospheric/Cloud indication classifies the relative sunlight by the presence or absence of a shadow around the vehicle. Shadows as well as diffuse lighting conditions when clouds are present can have a big impact on warning system performance. Clouds, hills, or large buildings may block the sun. The condition of the sunlight may explain warning system failures or highlight warning system performance. Analyze the entire video, not just during warnings. Volpe then uses the durations to determine the exposure time of the warning system to shadow and non-shadow conditions. This ensures a balanced, objective view of warning system performance (i.e., the system was barely tested in harsh sunlight).

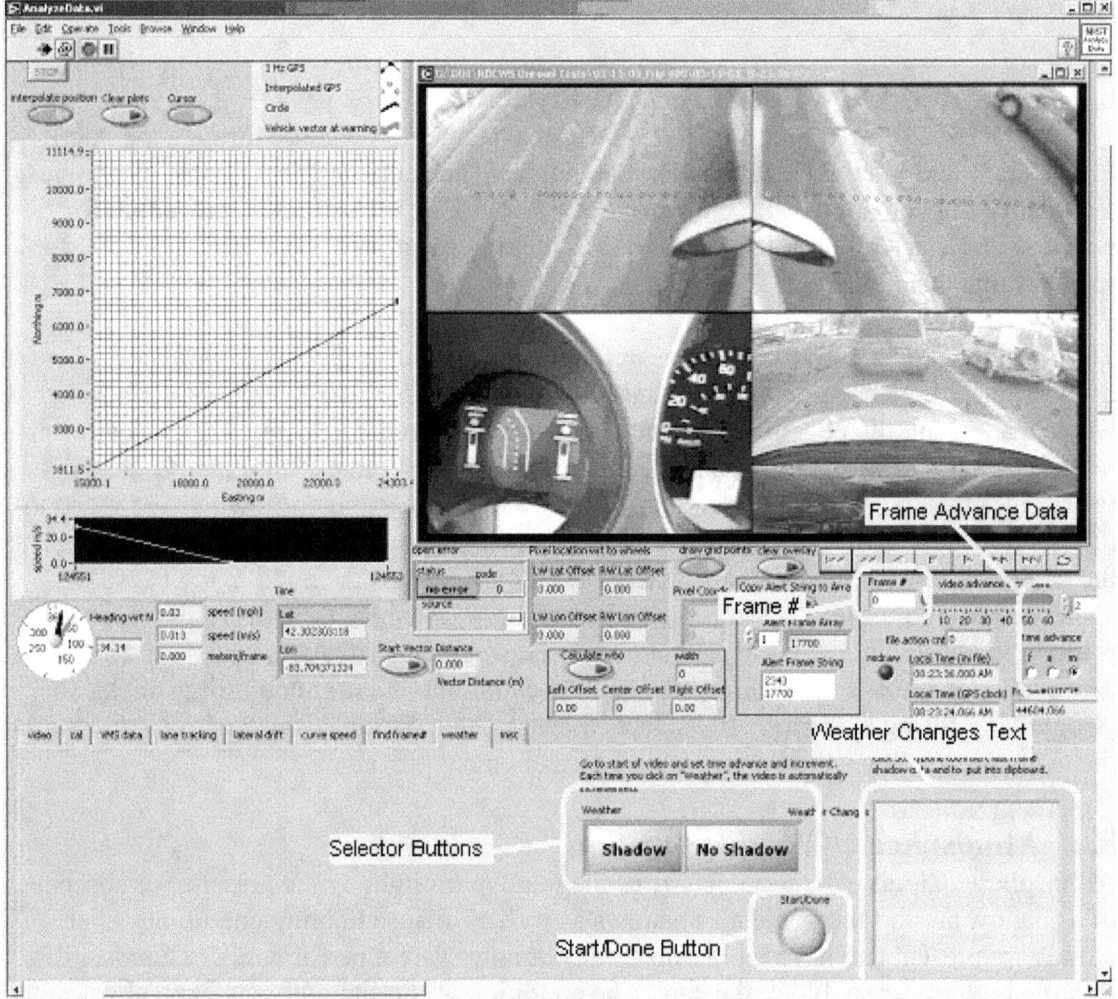

Figure 32. AnalyzeData "weather" Tab Display.

The weather analysis procedure uses the "weather" tab on the AnalyzeData VI. The analysis begins at frame zero and proceeds at 2 min intervals through the avi file. The analysis records the times when the shadow appears or disappears. The procedure requires a minimum amount of user motion and mouse clicks to collect the data.

1. Insert the number zero in the Frame # window.
2. Insert the number 2 in "video advance increment" and select "m" in "time advance".
3. Select the "weather" tab in the AnalyzeData.vi.
4. Click on the "Start/Done" button. The system will beep.
5. Check the avi image for the presence or absence of a shadow. The image at frame zero in Figure 32 shows a faint shadow around the vehicles ahead of the test vehicle. However there is no shadow around the test vehicle. Therefore, we record "no shadow" for this interval.
6. The avi display advances about 2 min when you select "Shadow" or "No Shadow".
7. Continue selecting Shadow or No Shadow as appropriate. Whenever there is a change, a new line will appear in the "Weather Changes" text window. The line

will include the GPS time, the local time, and a "1" for a shadow or a "0" when there is no shadow.
8. When the avi image no longer advances, click the "Start/Done" button to add the final condition to "Weather Changes". The program will automatically copy the "Weather Changes" text into the operating system's clipboard. (For example see Figure 33)
9. Paste the clipboard into the appropriate spreadsheet for future analysis.

Sec (from midnight)	Local Time	Shadow=1
68886.601	3:08:07 PM	x
68347.970	2:59:08 PM	1
69069.777	3:11:10 PM	0
69189.874	3:13:10 PM	1
69790.473	3:23:10 PM	0
69910.603	3:25:11 PM	1
70030.712	3:27:11 PM	0
70150.842	3:29:11 PM	1
71712.396	3:55:12 PM	0
71832.525	3:57:13 PM	1
71952.655	3:59:13 PM	0
72072.785	4:01:13 PM	1

Figure 33. Typical Weather Changes Summary.

9 Lane Marker Contrast

Lane marker contrast is the difference between the average intensity of the marker and the average intensity of the adjacent road surface. Figure 34 shows the control panel for making contrast measurements.

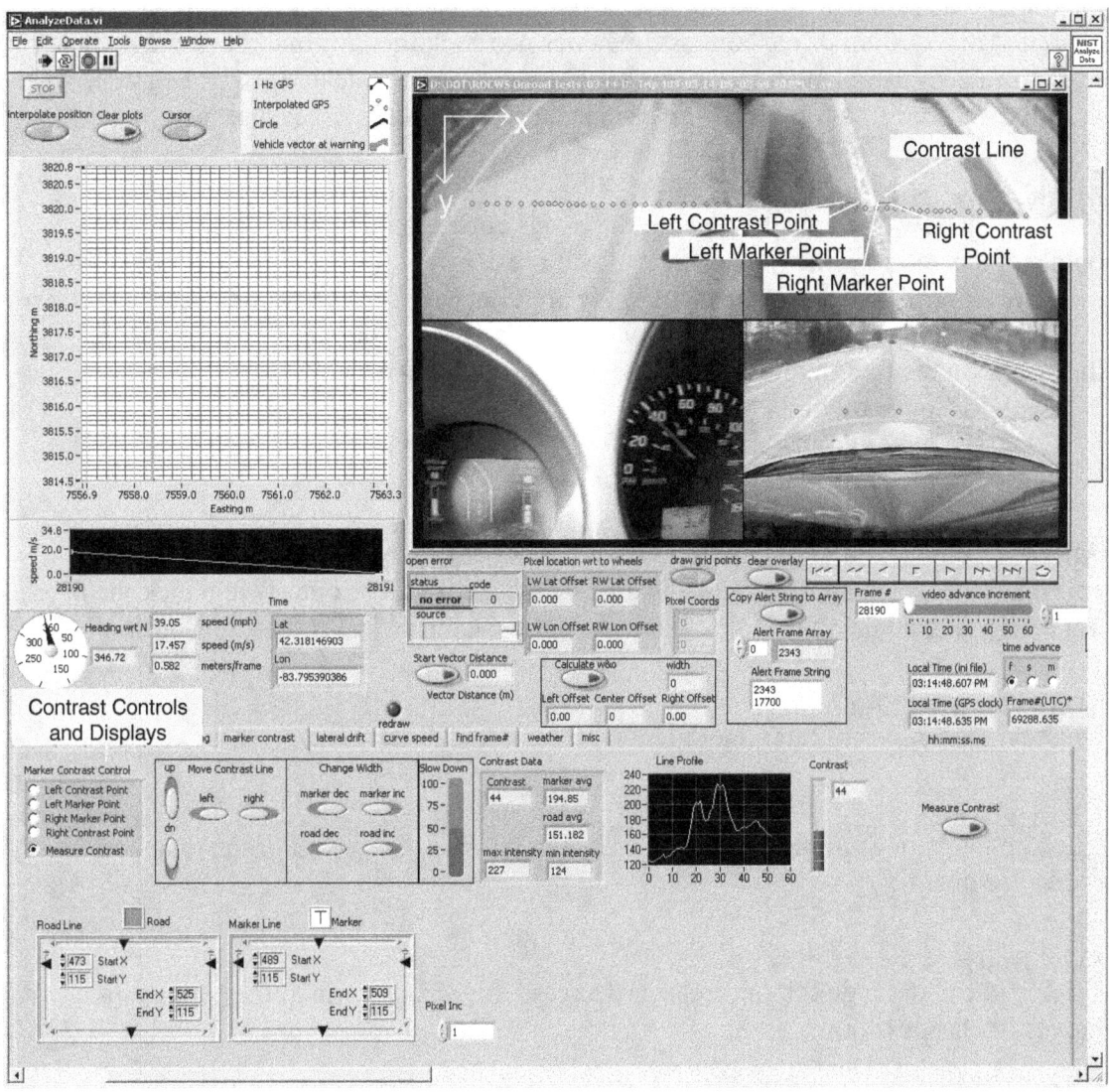

Figure 34. Control Panel for measuring lane marker contrast.

The "Contrast Line" in the video window shows the region in the image where the contrast measurement occurs. The line consists of two red segments, the left and the right road segments (if the segment is not visible, click on "Measure Contrast" to force a measurement and to cause the line to appear). The average of the pixel intensities of these segments produce the "road average". The region in between, which is transparent (no color), is the marker segment. The average of the pixel intensities of this segment produce the "marker average". The difference between the averages is the "contrast".

Note that for an RGB image, intensity is the average of the pixel's red, green and blue components. Yellow may stand out in an image, but its intensity can be "low" and correspondingly, the contrast may be low.

The following components, starting from the upper left component in the control window, are available to make a contrast measurement and to view the results. Contrast measurements automatically occur whenever the frame changes or the line changes. Note that the location of the contrast line does not track the lane marker as the marker shifts. An automatic lane finder can track the lane, but the tracker usually degrades when the contrast decreases; typically, the instance the user desires a measurement.

Marker Contrast Control
Use this window to manually click on the pixels corresponding to the start and stop points of the contrast line (however, it may be easier to move the current contrast line using the controls described below). There are four points to the contrast line, as shown in Figure 34). To start, click on "Left Contrast Point" (in "Marker Contrast Control") and then click on the four pixels in the image in the following order:
1. Left Contrast Point
2. Left Marker Point
3. Right Marker Point
4. Right Contrast Point

After each click, "Marker Contrast Control" advances to the next point. To redefine a point, just select the appropriate valued in "Marker Contrast Control" and pick the pixel. After selecting all the pixels, set "Marker Contrast Control" to "Measure Contrast".

Move Contrast Line
Click on the buttons in this panel to move the contrast line.

Change Width
Click on the buttons in this panel to increase or decrease the width of the road and the marker segments.

Slow Down
Click in the slider region to increase or decrease the speed at which the contrast line moves or changes width.

Contrast Data
1. marker avg – the average intensities of pixels corresponding to marker segment
2. road avg - the average intensities of pixels corresponding to road segment
3. Contrast = "marker avg" – "road avg"
4. max intensity – the maximum pixel intensity over the road and marker. Used to fix the scale of "Line Profile" so that it remains constant while viewing a video.
5. min intensity – the minimum pixel intensity over the road and marker. Also used to fix the scale of "Line Profile".

Line Profile
This is a graph of intensities for each pixel in the contrast line.

Contrast
This is a bar graph of the contrast (same value in "Contrast Data").

Measure Contrast
Use this to force a contrast measurement in cases when one does not automatically occur.

Road Line
These are the coordinates of the Road line. Note these coordinates are the same as "Left Contrast Point" and "Right Contrast Point" described above. The split into two segments (left and right) uses the coordinates of the marker line.

Road
This is the color (currently red) to overlay for the road segment. Click in the box to bring up a menu of alternative colors.

Marker Line
These are the coordinates of the marker line. Note these coordinates are the same as "Left Marker Point" and "Right Marker Point" described above.

Marker
This is the color (currently transparent) to overlay for the marker segment. Click in the box to bring up a menu of alternative colors.

Pixel Inc
Another way to adjust speed when moving or resizing lines.

10 MapPoint Map Display

If the host computer has the Microsoft MapPoint application, the Lateral Drift Analysis software displays the GPS position on an external map. The software adds a pin at the coordinates of the frames displayed. If an alert frame occurred (as recorded in the Alert Frame Array) between the last pin and the current frame, the map plots a red pin, otherwise a blue dot. Figure 35 shows a typical run near Ann Arbor, MI. The map shows a series of alerts along the lakeshore, no alerts in the downtown area, and a GPS jump at the intersection on the western side. NIST intends to integrate more functions with the maps in subsequent versions of the evaluation software.

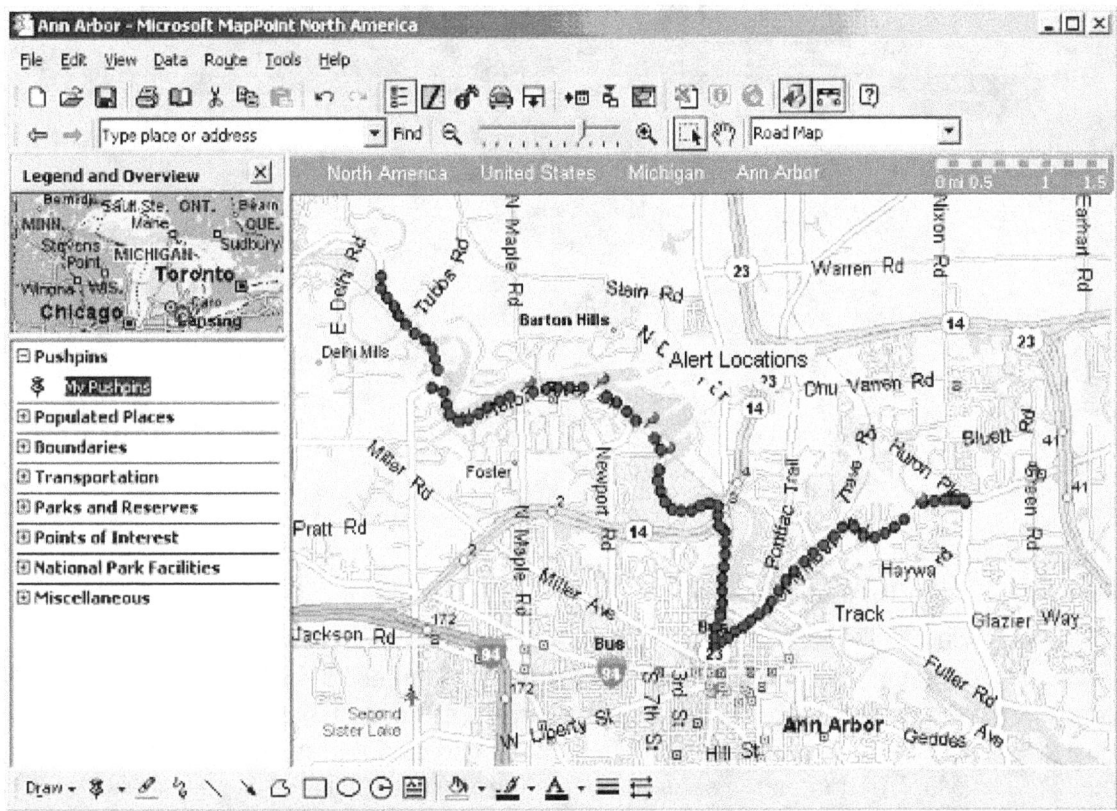

Figure 35. Map Point Display.

The analysis software requires the computer to have a map file saved on the disk. The operator specifies this file when MapPoint starts up. The operator may change the map file by using the "Select New Map" switch in the external map controls. If "Select New Map" is active, the software will ask for a new file the next time the operator starts a new track, otherwise the display reuses the previous map.

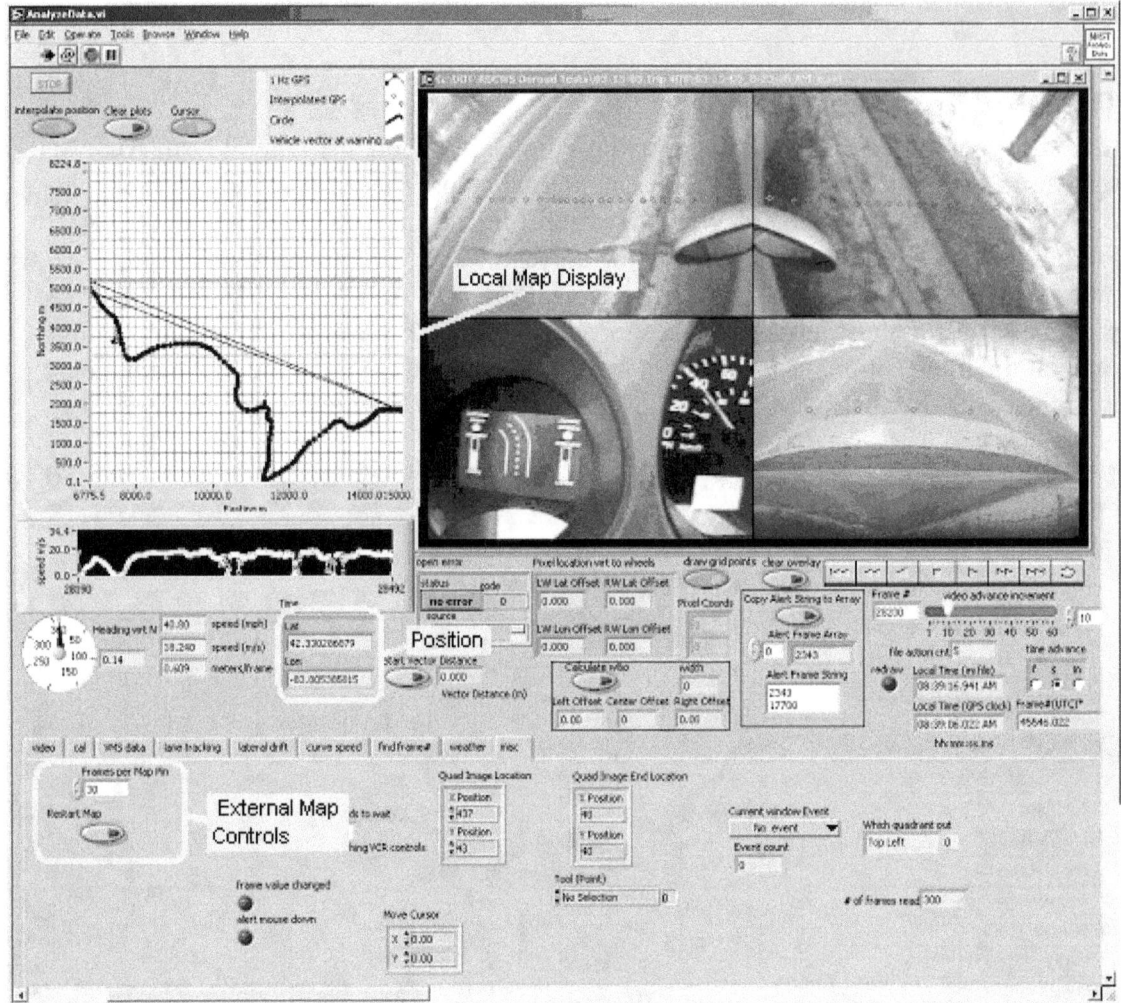

Figure 36. Map Interface.

The operator uses the "Restart Map" button in the external map controls to start a new track. The operator may first save the map for future reference by using the MapPoint controls. The operator may also clear the track with the delete function under "My Pushpins" on the MapPoint interface. After saving and clearing, the operator activates the "Restart Map" button on the external map controls on the data analyzing software display (see Figure 36).

The operator may limit the number of pins by setting the frames/pin selector under the "misc" tab (see Figure 36). The frames/pin selector instructs the software to add a pin only after the specified number of frames has passed. When the frame index rate is greater then the frames/pin selector, the software adds only one pin per index. The default rate is 30 frames per pin, which equals 1 pin per second of continuous running. The operator sizes, centers, and clears the map as required using the MapPoint controls.

References

[1] Request for Applications for Cooperative Agreement DTFH61-01-X-00053, A Intelligent Vehicle Initiative Road Departure Crash Warning Field Operational Test, April 11, 2001, http://www.eps.gov/EPSData/DOT/Synopses/30/DTFH61-01-X-00053/DTFH61-01-X-00053.rtf

[2] Bruce Wilson, Jonathan Koopman, Sandor Szabo, "Roadway Departure Crash Warning System Performance and Capability Testing", March 7, 2005.

Appendix A Adobe Premier Notes

This section contains notes describing tips and quirks uncovered using Adobe Premier 6.0.

A.1 Differences in file lengths

The following is an inconsistency discovered in the length of an avi file when viewed in Premier and in Labview. Differences in file length cause timing mismatches between the video and the inserted GPS/time stamp. By the end of a long video file, the GPS data indicating the position and velocity of the vehicle does not match the view of the vehicle's motion in the video (vehicle appears to stop yet the GPS position indicates the vehicle is still moving). The solution is to ensure the compression frame rate matches the capture frame rate. The following example describes the problem in more detail.

When looking at the clip "05-28-03 LDW Barriers_c.avi" in the Premier timeline, the monitor window indicates the last frame time as 00:41:43:02 (hours:minutes:seconds:frames). This translates to 75092 frames.

However, when examining the properties of the clip (under the Premier file menu), the duration is 00:41:45:18 and the number of frames is 75168. In addition, when looking at the file in LabView, the number of frames is 75168.

There is a difference of 75 frames (approximately 2.5 s) between the timeline length and the properties/Labview length. The DV capture frame rate is 29.97 frames/s. The AVI file properties (viewed by Labview) shows 30 frames/s, which was the setting used during compression (all captured DV is compressed to save file space and improve throughput). To prevent a mismatch in timing between video frames and GPS data associated with each video frame, make sure the compression frame rate matches the capture frame rate.

Appendix B Creating a DVD using DVDit!

The following section describes how to create a DVD using DVDit (www.sonic.com), which is bundled with Premier. The first step is to use Premier to create an m2v (video) file and a wav (audio) file (see the Premier help for creating these files).

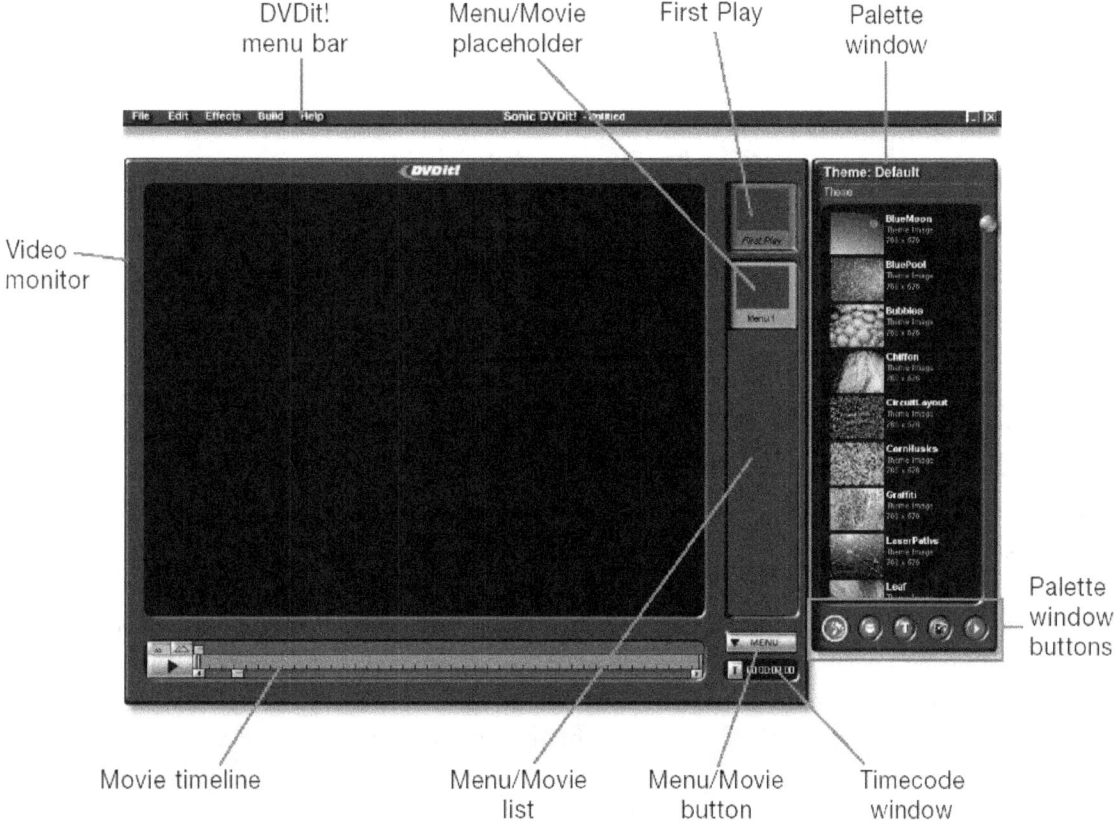

Figure 37. The main window in DVDit

The next step is to import the files into DVDit and to build the DVD menus. The steps below refer to Figure 37.

1. Use default theme

2. Create menus
   - Select menu under Menu/Movie button
   - Select Background under Pallet window buttons
   - Drag a background into Menu/Movie placeholder in Menu/Movie list.
   - Edit menu name, e.g., "main menu"

3. Create menu entries (buttons or text).
   - Select menu under Menu/Movie button
   - Select appropriate menu (may already be visible in Video monitor)

- Select either Button (button 2) or Text (button 3) under Pallet window buttons
- Drag item onto menu and edit text appropriately

4. Import media
   - Select Media (button 4) under Pallet window buttons
   - Right click in window and select "add files to theme"
   - In pop-up browser, highlight .m2v and .wav files (should have same name) and click open. This will insert media into Pallet window

5. Select first play
   - Select movie under Menu/Movie button
   - Drag media (both video and audio files) onto First Play placeholder

6. Link media to menu buttons
   - Make sure menu is visible (select Menu under Menu/Movie button)
   - Drag media (mv2 and wav files) onto button

The following steps describe how to burn the DVD. These steps appear at http://support.sonic.com/kb/default.asp?type=desktop&product=dvdit

Title
Building a DVD Folder to your hard drive using DVDit! and then burning a DVD.

DVDit can burn a project right to a DVD. However, creating a volume on a drive enables you to make additional copies without having to have the original DVD to make a copy. Attempting to create a volume on the disk failed with unknown error. The following procedure did work. Before you begin, create two new folders on the top level of a local hard drive. The hard drive must NOT be part of a RAID or network configuration. Name one folder "DVDit! Temp" and the other the name of you project, e.g. "Wedding Video". When you finish creating your session, build your session to your hard drive. You can do this by going to your Build Menu-->Make DVD Folder.

1) GENERAL TAB

Source: Current Project.
Output Options: Should be blank.
Path: Top Level of hard disk. NOTE: Make sure that the drive you select is local to your machine and not a RAID or network drive.

2) ADVANCED TAB

uncheck all boxes

Path: Select your DVDit! Temp folder. Browse to this folder in your Path dialog and select it just in case you decide later down the line to save your cached files. This will make them easy to locate for later use or deletion if they become corrupt.

3) Click OK. The build process begins.

When finished, burn the Volume to DVD. Go to the drive you built the project on and place the AUDIO_TS and VIDEO_TS folders into the second folder you created for your project title (e.g. "Wedding Video"). This prevents errors in DVDit! from accidentally burning data files with your video files.

From here, you have two options: you can either burn your DVD using DVDit! Or you can burn your DVD using a third party data burning application such Roxio or Prassi.

NOTE: BE SURE THAT YOUR BURNING APPLICATION SUPPORTS YOUR BURNER AND YOUR DVD MEDIA TYPE!

OPTION 1--Using DVDit!:

1) Go into your DVDit! Build Menu-->Make DVD Disc.

2) GENERAL TAB

Source: DVD Volume. Click the browse button. Locate your session folder (e.g. "Wedding Video") and open it so you can see the AUDIO_TS and VIDEO_TS folders. These should be the only folders.

Output Options: Check this ONLY if you are creating a cDVD.

Recorder: Select your recorder

Number of Copies: 1

Options: Test and Create disc

3) Click OK.

Burning begins.

OPTION 2--Using a third Party Application

1) Choose to create a Data disc in your burning application

2) Drag the AuDIO_TS and VIDEO_TS folders into your to-be-burned or data field. Be sure you have dragged the folders ONLY and no other files. These are the only two items (and their contents) to burn.

3) You may want to run in simulation mode if you haven't done this process before to make sure you have the proper buffer speed and RAM allocated to your machine for a clean burn. If it works, then burn the disc for real.

The following error occurred in an attempt to create a DVD.

The solution below appears on the Sonic website. Actually, this problem occurs when "current project" is not selected in Build->Make DVD folder->general->Source.

Title
DVDERR.-16019

Solution
This is bug that we are currently working on fixing in a new release in MyDVD/DVDit!. So far, we have been able to determine that it may have to do with where your files are placed on the hard drive and where they have come from. Usually, this happens when the audio_ts and video_ts folders from a pre-made DVD (either Hollywood or a consumer made) are dropped onto the hard drive and the user tries to replicate them into another DVD or "back up". Also, the error can be caused due to renaming the file, using non-alphanumeric characters (8 or less), or incorrect path settings as when the files has been moved while the session is in use.

You may also want to be sure that you are building your session to the top level of your hard drive as well. Please follow the links on this page on how to do this.

Appendix C Extracting GPS data from video using VMS and MediaMapper

The VMS 300 is a device made by Redhen Systems (www.redhensystems.com) that consists of a DGPS receiver and electronics to convert the DGPS data into an audio signal for insertion on videotape. The VMS is also required to read the DGPS from the videotape.

MediaMapper is the software provided by Redhen Systems to read the GPS data off the video tape and perform differential correction using EZDIFF (from within MediaMapper). MediaMapper is not required to write GPS data to the tape. The VMS does this automatically when the VMS is on and connected to the VCR.

MediaMapper also contains EZDIFF to correct the GPS data if the VMS is running in differential mode (i.e., saves all satellite data).

The corrected data is stored in an ASCII file. The corrected data allows the user to determine vehicle position and speed in the avi file at any point in time. The data is available once per second.

The following sections describe the steps for creating the corrected GPS data file.

C.1 VMS SetUp

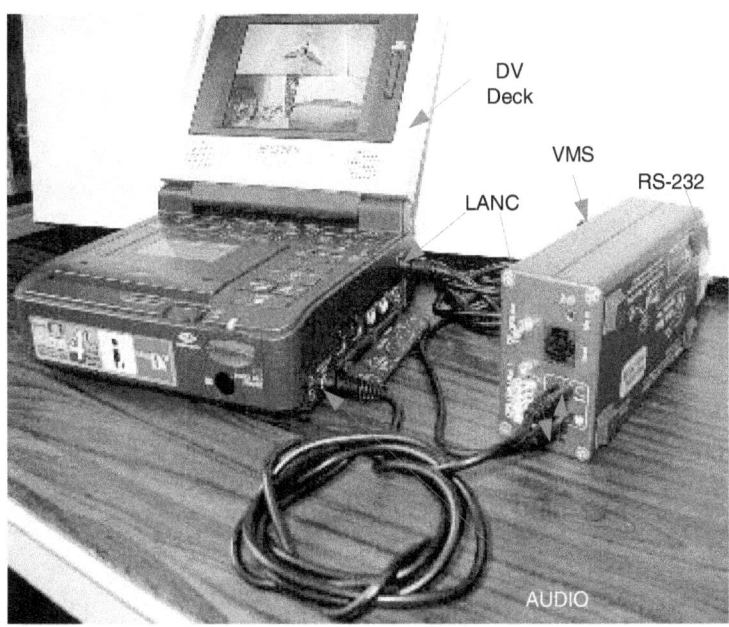

Figure 38. Connecting the VMS to a VCR.

The VMS serves as an interface between MediaMapper and the VCR. There are two cables between the VMS and the VCR and one cable between the VMS and the host computer for MediaMapper. The VMS requires an audio connection to the VCR's audio/video jack so that it can read the GPS data off the audio track (the GPS data sounds like a hiss and a beep every second). A LANC interface allows the VMS to send VCR controls commands (play, stop, etc.) to the VCR and to read the tape counter time from the VCR. An RS-232 cable connects to the back of the VMS to the host computer's COM port. MediaMapper uses the serial interface to read the GPS and tape counter data and to sends tape commands (stop, play, etc.) to the VCR (via the VMS Lanc interface).

C.2 Extracting GPS data with MediaMapper

The goal of this step is to extract the GPS data that corresponds to a video clip. The GPS data file and video clip have the same root name.

In MediaMapper, indexing a video means extracting the GPS data. The following steps are performed within MediaMapper. NOTE: MediaMapper only displays GPS points (i.e., the red points on the map) when indexing. When playing a tape you will only see an arrow move across the map without any trail of GPS points showing the previous positions.

1. File->Add GPS Index

2. Pick a name that corresponds to video file, e.g., 05-28-03 LDW Barriers_c. The program saves the file in the user/My Documents/Map Library by default. This is OK for now since the user will later convert the file into an Excel file and store it in the same directory as the video file.

3. Index parameters:
   Use same name as video file
   The start time for the example clip is:
   00;15;48;24
   The number 24 indicates the frame count. MediaMapper time resolution is only 1 s. Start the capture a few seconds early, you can always edit out the unwanted times later using Excel.

   MediaMapper always rewinds to the beginning of the tape so do not bother moving the tape to the capture point. In addition, you cannot set the stop point, so set your stopwatch to the duration of the clip.

   While indexing, you should be able to see the GPS data and the serial data if you enable the displays (satellite and cable icon buttons below map to right).

   Noted Problems:
1. The play button in MediaMapper would not cause VCR to play. Fix: rewind first, tape may have been at the end.

2. The GPS data does not display. Fix: The audio volume may be too high (recall the GPS data is stored as an analog signal on the audio channel). Lower to around 1/3 volume.

3. GPS drop out. Fix: Raise volume slightly. Dropouts still occurred. Went to tools->options->VMS setup->Camera Setup and checked box indicating RC time code or digital time (note that equipped with microphone power was also checked). No more dropouts occurred.

4. When indexing an entire tape, if tape ends, MediaMapper sometimes aborts with an access violation error (see Figure 39). Fix: The files are still useable. Go to Map Library directory and double click on the name of the map you created, for example "03-15-05_8-23-36 AM_c.vtc". When the map comes up, it contains no GPS points. To add the GPS data, go to File->Add Layers, go inside the index folder, and click on "03-15-05_8-23-36 AM_c.tab" (.tab files contain all the layers associated with a map file). Then you should see the trail of GPS points and can run post processing using EZDiff.

Figure 39. Mediamapper sometimes hangs while indexing GPS and the video ends.

C.3 Post process GPS data using EZDiff

EZDiff is a software tool packaged with MediaMapper. The tool downloads base station corrections used to correct GPS data from a roving receiver. The base stations are part of the National Geodetic Survey's Continuously Operating Reference Stations (CORS). To run EZDiff:

1. Select Tools->Differential Correction

Retry if there is an FTP error getting on the CORS site. Below is an example processing-log from a running of EZDiff.

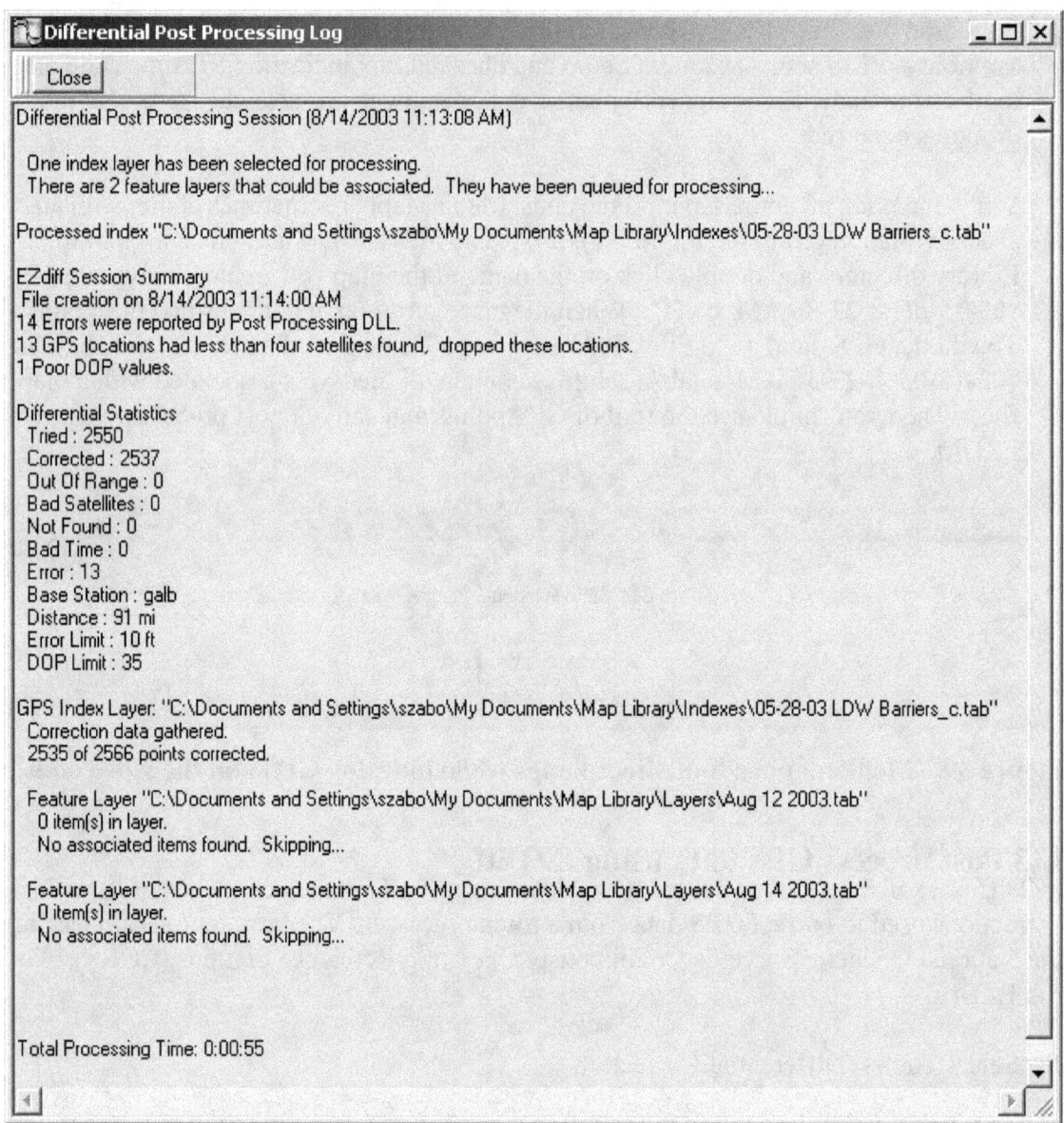

After EZDiff performs corrections, it produces a DBF (database file) called 05-28-03 LDW Barriers_c.DBF (in C:\Documents and Settings\szabo\My Documents\Map Library\Indexes), which can be opened with Excel.

Noted Problems:

EZDiff produced the following error on 5/17/2004.

****

FTP Session Log File: 5/17/2004 2:37:16 PM EZdiff V2.0.0.18

Connecting to [www.ngs.noaa.gov].

Setting mode to image (binary).

Changing directory to [/cors/rinex].
Successful directory change to [/cors/rinex].

Changing directory to [2004/118].
Successful directory change to [/cors/rinex/2004/118].

Get directory listing of [/cors/rinex/2004/118].
Successful download of directory listing for [/cors/rinex/2004/118].

Changing directory to [galb].

Get directory listing of [/cors/rinex/2004/118/galb].

Changing directory to [/cors/rinex/2004/118].
Successful directory change to [/cors/rinex/2004/118].

Changing directory to [stkr].

Get directory listing of [/cors/rinex/2004/118/stkr].

Changing directory to [/cors/rinex/2004/118].
Successful directory change to [/cors/rinex/2004/118].

Changing directory to [erla].

Get directory listing of [/cors/rinex/2004/118/erla].
Error: Rover data for today, hour [19] is too new to found on CORS FTP site.
Try again later.

Errors were found during FTP session.

Fix:
1. Went in to options and Selected nearest CORS site and it returned DET1. Received a similar error message as above ("to new to found on CORS FTP site">
2. Downloaded New List – still got error
3. Clicked on Select CORS site and chose det1. This worked.

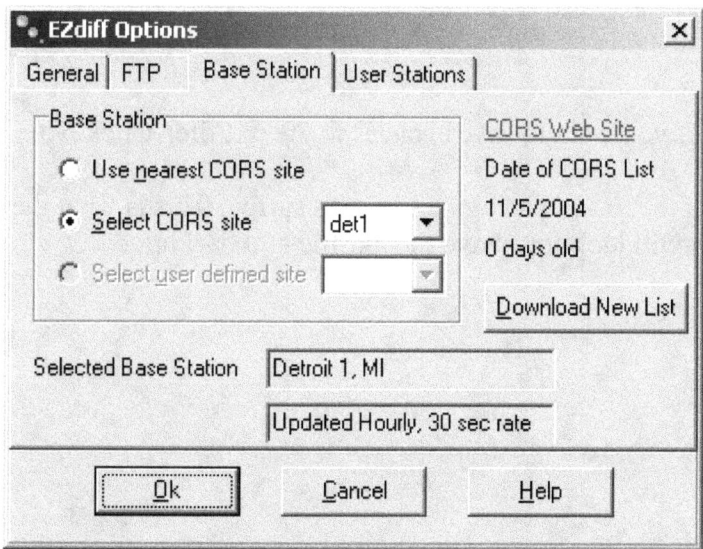

C.4 Create a jpg of GPS Points
Follow the steps below to create a jpg of the map showing the GPS points read from the videotape.

1. Use zoom and hand to draw a window on the portion of the map containing the GPS points.
2. Select window using rectangle tool.
3. Select file->export->graphic.
4. Pick save type, for example, jpg.
5. Choose folder to save the jpg.

C.5 Convert to LabView format using Excel
The next step converts the dbf file into a vms file, which is the format used in Labview.

1. Open a .DBF file using Excel. The .DBF file appears in My Documents\Map Library\Indexes.

2. Find and replace "Post Diff" with PostDiff (LabView string processing routines don't like spaces inside a string).

3. Remove last column called WAYPOINT if there is nothing in the column.

4. Save as .txt file (pull down Save as type: menu and select "Text (tab delimited) (*.txt)") with same name as video(.avi) file in the same directory as the video file

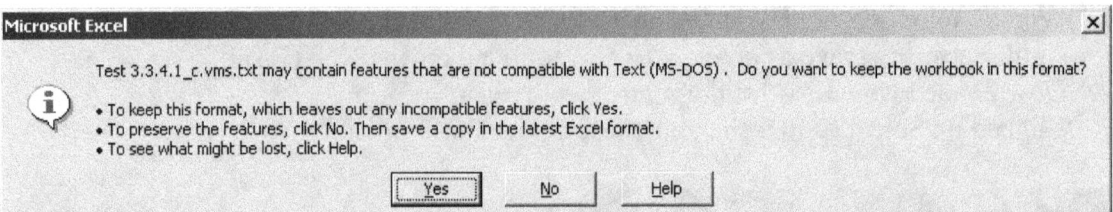

Click Yes

5. When you quit, Excel will ask if you want to save changes to the .txt file. Click No.

6. Go to directory, make a copy of the .txt file and give it a .vms suffix. This is what the LabView analyze data program will look for. Save the .txt file as a backup.

Appendix D Nonlinear Camera Calibration

A non-linear calibration technique implemented in AnalyzeData makes measurements over a large area. Figure 40 illustrates aspects of the calibration process. A calibration grid is required to calibrate the camera. The operator examines an image of the grid and identifies each grid point (Figure 41). The operator also enters the actual distances between the grid points, in meters. A nonlinear calibration routine runs and produces the transformation parameters. The parameters remove distortions in the image (Figure 42). The procedure requires a large number of grid points, which requires substantial time for setup and takedown, and for processing. When possible, use the calibration stick method.

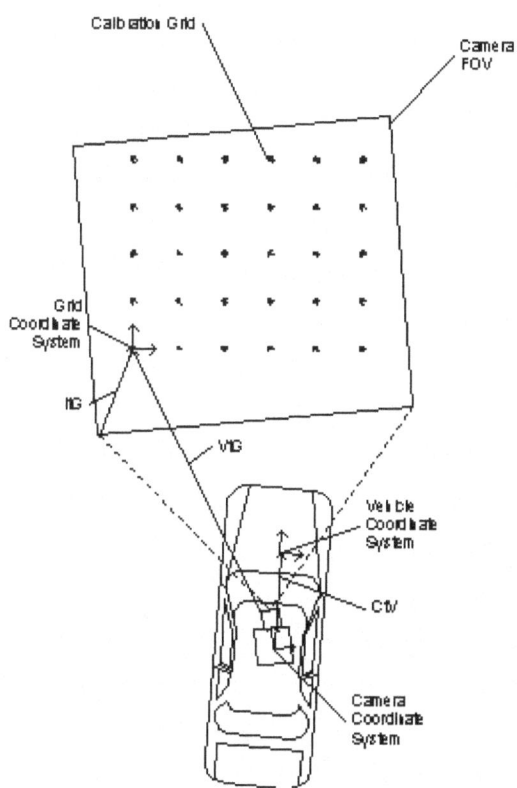

Figure 40. Camera calibration coordinate systems.

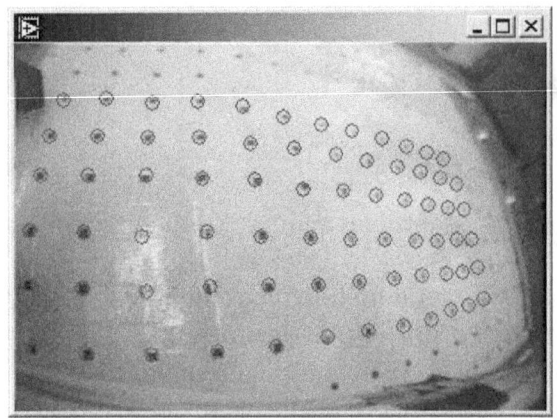 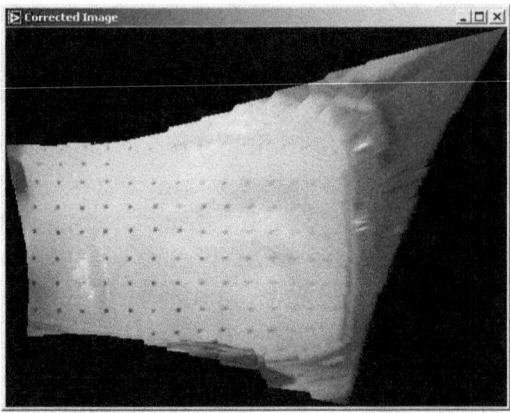

Figure 41. Nonlinear calibration grid with identified grid points (red circles).

Figure 42. Nonlinear distortions removed after calibration.

Appendix E UTM Conversions

Universal Transverse Mercator (UTM) coordinates use a family of projections based on the Transverse Mercator projection. The projection consists of an ellipsoid divided into 60 longitudinal zones of 6° each. The X value, called the Easting, has a value of 500,000 meters at the central meridian of each zone. The Y value, called the Northing, has a value of 0 meters at the equator for the northern hemisphere, increasing toward the north pole, and a value of 10,000,000 m at the equator for the southern hemisphere, decreasing toward the south pole.

The conversion routines used in the IMS use the Geographic Translator (GEOTRANS) software package. The source code is available from the National Geospatial-Intelligence Agency at http://earth-info.nga.mil/GandG/geotrans/.

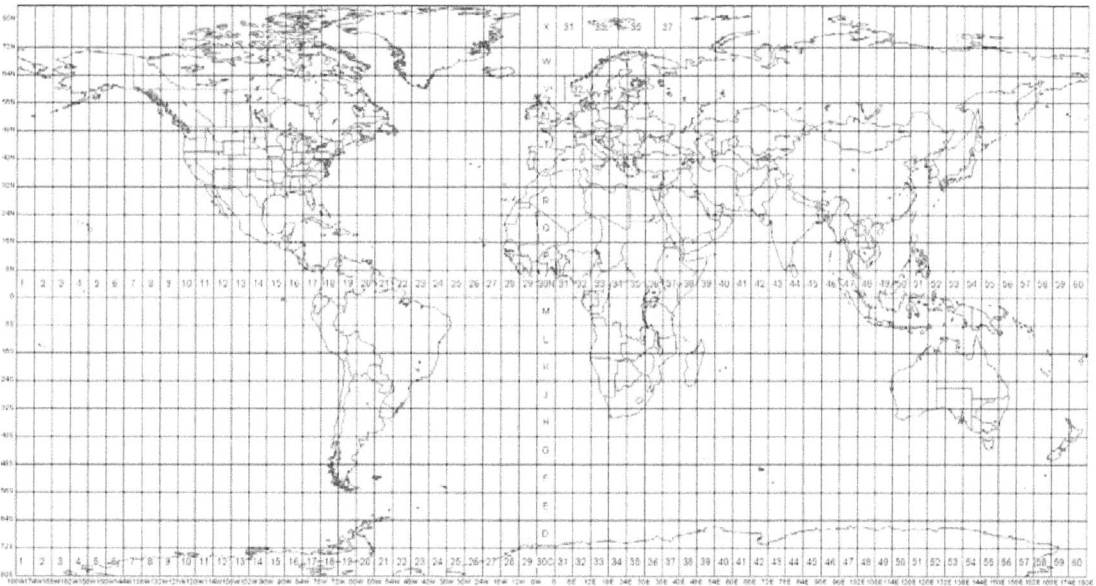

Figure 43. UTM and Military Grid Reference System zones. The UTM goes from 1-60 for each hemisphere. The MGRS divides the zone laterally using letters C-X.